Lecture Notes in Computer Science

Lecture Notes in Artificial Intelligence 14951

Founding Editor

Jörg Siekmann

The series Lecture Notes in Artificial Intelligence (LNAI) was established in 1988 as a topical subseries of LNCS devoted to artificial intelligence.

The series publishes state-of-the-art research results at a high level. As with the LNCS mother series, the mission of the series is to serve the international R & D community by providing an invaluable service, mainly focused on the publication of conference and workshop proceedings and postproceedings.

Kristinn R. Thórisson · Peter Isaev ·
Arash Sheikhlar
Editors

Artificial General Intelligence

17th International Conference, AGI 2024
Seattle, WA, USA, August 13–16, 2024
Proceedings

Editors
Kristinn R. Thórisson
CADIA
Reykjavik University
Reykjavík, Iceland

Peter Isaev
Temple University
Philadelphia, PA, USA

Arash Sheikhlar
CADIA
Reykjavík University
Reykjavík, Iceland

ISSN 0302-9743 ISSN 1611-3349 (electronic)
Lecture Notes in Artificial Intelligence
ISBN 978-3-031-65571-5 ISBN 978-3-031-65572-2 (eBook)
https://doi.org/10.1007/978-3-031-65572-2

LNCS Sublibrary: SL7 – Artificial Intelligence

This Springer imprint is published by the registered company Springer Nature Switzerland AG
The registered company address is: Gewerbestrasse 11, 6330 Cham, Switzerland

Preface

This volume contains the papers presented at the 17th Conference on Artificial General Intelligence (AGI 2024), held in Seattle, USA, August 13–16, 2024. There were 55 submissions, each single-blind reviewed by at least two Program Committee members. Out of 25 papers accepted (45% acceptance rate) 18 papers were selected for oral presentation, and 7 papers for poster presentation. Many excellent keynotes were given by researchers from both academia and industry. The conference also featured tutorials and workshops.

Even though the idea of creating machines that possess the magic of intelligence dates back far in history, little progress was made on how this could be achieved until the advent of electronics at the turn of the 20th century. Electronic circuits gave early pioneers of what is now called "artificial intelligence" (AI) a practical framework to think about how information states could be created and manipulated to achieve artificial classification functions and control. With the advent of the transistor, in 1947, the work kicked into high gear, and even more so with the advent of integrated circuits.

Before computing machinery, resting on the electronics revolution, all ideas for making "artificial humans" were hopelessly simplistic. This did not stop the early pioneers of the field being optimistic, however, predicting the imminent arrival of AGI. This tradition has lived on to the present day – many researchers and laypeople alike now feel that human-level artificial intelligence, possibly even superintelligence, is just around the corner, echoing predictions from every preceding decade since the 1950s. Like in the decades prior, predictions about the rise of generally intelligent machines are pushed back at a rate of approximately one year per year.

But is AGI doomed to always be 10 years away? After all, haven't we been making significant progress since the 1950s? What are the missing ingredients for the recipe of human-like intelligence?

The principal process behind intelligence is thought: Thought is the driving force of human society, powering invention, creativity, communication, and everything else that humans do. In the early days of AI – the term was coined in 1956 – quite some space was given to discussions of what 'thinking' is, in papers and tech reports, as well as related concepts such as understanding, reasoning and meaning, although the proposed ideas were too technologically simplistic to make a dent in these challenging phenomena. In the 1970s and 80s such terms and several other related ones fell out of favor as central subjects of AI research, however, with work on formal, provable systems filling the gap. As the limitations of strictly formal methods for addressing questions about real intelligence begin to show, these topics are being reclaimed from philosophy and increasingly given new status in scientific discourse.

AI serves two masters, science and engineering. Both employ the scientific method for doing research, with the aim to create new knowledge, but they have different goals: In science we seek to uncover an unknown model; in engineering we aim to meet the specifications of a given model, 'the model' being a detailed explanation of how things work, typically referred to as 'theory.' For AI researchers this duality is a feature, not a bug: A complex dynamic phenomenon like intelligence needs powerful tools, which engineering provides in the form of models that can be run on computers and compared to their natural equivalents. But the biggest contribution to scientific knowledge that AI could provide is a comprehensive theory of intelligence. No other field has "intelligence" in its title; no other field is as well poised to make this contribution. No real progress can be made without it.

As of yet, no generally accepted comprehensive theory of intelligence or thought exists. The rate of progress in AI will continue to be determined by the amount and quality of basic research. For the past three decades much of AI research has been conducted under a rubric of what may be called 'applied basic research,' where the main topics being focused on are selected in light of what seems promising in the near future – preferably this year or next year. Artificial neural nets, for instance, and technologies of their ilk, are similar neither in operation nor in structure to natural neural networks, nor do they demonstrably capture in some way the information generated in natural brains. Relevance to scientific investigations of how intelligence works cannot therefore possibly be the main reason for why they feature so prominently in today's AI research.

An underlying assumption in AI research is that the phenomenon of intelligence can be captured in large part or fully by how information is generated and used. The field has certainly made significant progress on its subject matter in the past 70 years. Yet there is little doubt in the minds of most researchers attending the AGI conference that today's AI is still very different in many ways – and significantly inferior – to human intelligence.

As the name states, the main subject of the AGI Conference is intelligence in general, specifically *general intelligence*. The theme of this year's conference is "Understanding AGI." Many central concepts of general intelligence, including *thought*, *understanding*, *meaning*, *creativity*, *insight*, *reasoning*, *autonomy*, *attention* and *control*, can be found in some form or other in the present volume.

If 'intelligence' is that special something that enables a machine or an animal to achieve a specific goal with respect to some particular phenomenon, in light of uncertainty and incomplete knowledge, and get better at that over time, 'general intelligence' is the ability to do so for many sets of phenomena. Such systems should be able to learn in multiple ways, improve their knowledge over time, handle multiple levels of detail and time, and have some sort of common sense, which involves forming a coherent logical picture of things in spite of incomplete knowledge. Having general intelligence also means being able to figure things out in light of impending deadlines, misleading information, uncertainty and numerous other challenges, knowing when experimentation and various forms of analogies can be used to make sense of one's experience, rule out incorrect explanations, consider and reconsider what is possible and impossible, identify contradictions, exposing what one doesn't know – and think about thinking itself. These are some of the topics that constitute what AGI research seeks to uncover the fundamental mechanisms for – this year – and in the coming decades.

We thank all the Program Committee members for their excellent work and congratulate the authors of the published papers for moving the field of AI forward on challenging and critical topics. We also thank all of our contributors and collaborators:

Keynote speakers, committee members and organizers of tutorials, workshops and panel sessions.

Lastly, we thank our sponsors: the Artificial General Intelligence Society, Springer Nature, SingularityNET, TrueAGI and OpenCog Foundation.

June 2024

Kristinn R. Thórisson
Peter Isaev
Arash Sheikhlar
Matt Iklé
Ben Goertzel

Organization

Conference Chair

Kristinn R. Thórisson	Reykjavik University, Iceland

Organizing Committee Chair

Matt Iklé	SingularityNET, USA

Keynote Speaker Invitations, Conference Themes

Ben Goertzel	SingularityNET, USA

Program Committee

Pulin Agrawal	Pennsylvania State University, USA
Salem Benferhat	Cril, CNRS UMR8188, Université d'Artois, France
Michael Bennett	Australian National University, Australia
Adrian Borucki	Aperto, Poland
Cristiano Castelfranchi	Institute of Cognitive Sciences and Technologies, CNR, Italy
Antonio Chella	Università di Palermo, Italy
Berick Cook	SingularityNET, USA
Blerim Emruli	SICS Swedish ICT AB, Sweden
Nil Geisweiller	SingularityNET Foundation, Bulgaria
Michael Giancola	Rensselaer Polytechnic Institute, USA
Naveen Sundar Govindarajulu	Rensselaer Polytechnic Institute, USA
Christian Hahm	Temple University, USA
Patrick Hammer	Temple University, USA
Matt Iklé	SingularityNET, USA
Peter Isaev (Chair)	Temple University, USA
Nino Ivanov	Oracle, Austria
Min Jiang	Xiamen University, China
Robert Johannson	Stockholm University, Sweden

Garrett Katz	Syracuse University, USA
Anton Kolonin	Aigents, Russia
Xiang Li	Temple University, USA
Kai Liu	Bohai University, China
Tony Lofthouse	Stockholm University, Sweden
Haley Lowy	SingularityNET, USA
Brett Martensen	Adaptron Inc., Canada
Yoshihiro Maruyama	Kyoto University, Japan
Michael S. P. Miller	SubThought Corporation, USA
Alidoust Mohammadreza	Islamic Azad University of Science and Research of Tehran, Iran
James Oswald	Rensselaer Polytechnic Institute, USA
Alexey Potapov	SingularityNET, Russia
Rafal Rzepka	Hokkaido University, Japan
Oleg Scherbakov	ITMO, Russia
Hedra Seid	SingularityNET, USA
Arash Sheikhlar (Co-chair)	Reykjavik University, Iceland
Maxim Tarasov	Temple University, USA
Kristinn R. Thórisson	Reykjavik University, Iceland
Mario Verdicchio	Università degli Studi di Bergamo, Italy
Peter Voss	Aigo.ai, USA
Pei Wang	Temple University, USA
Robert Wünsche	TU Dresden, Germany
Bowen Xu	Temple University, USA
Roman Yampolskiy	University of Louisville, USA
Eyob Yirdaw	iCog Labs, Ethiopia

Contents

Generative AI Can Be Creative Too

Pulin Agrawal[1(✉)], Arpan Yagnik[1], and Daqi Dong[2]

[1] The Pennsylvania State University, The Behrend College, Erie, PA 16563, USA
pagrawal@psu.edu
[2] Memphis, TN 38017, USA

Abstract. Large Language Models (LLMs) have significantly influenced everyday computational tasks and the pursuit of Artificial General Intelligence (AGI). However, their creativity is limited by the conventional data they learn from, particularly lacking in novelty. To enhance creativity in LLMs, this paper introduces an innovative approach using the Learning Intelligent Decision Agent (LIDA) cognitive architecture. We describe and implement a multimodal vector embeddings-based LIDA in this paper. A LIDA agent from this implementation is used to demonstrate our proposition to make generative AI more creative, specifically making it more novel. By leveraging episodic memory and attention, the LIDA-based agent can relate memories of recent unrelated events to solve current problems with novelty. Our approach incorporates a neuro-symbolic implementation of a LIDA agent that assists in generating creative ideas while illuminating a prompting technique for LLMs to make them more creative. Comparing responses from a baseline LLM and our LIDA-enhanced agent indicates an improvement in the novelty of the ideas generated.

Keywords: Cognitive Architectures · Generative AI · Creativity · Large Language Models · Novelty · LIDA · Prompt Engineering

1 Introduction

Large Language Models (LLMs) have revolutionized how people engage with their everyday tasks. Generative models are trained to learn from a dataset, discovering structures and patterns that constitute that data using the attention mechanism [1], to generate possible next item in the dataset. Large Language Models like GPT-4 [2], are applications of generative AI in the context of natural language processing tasks [3] and an attempt to model artificial general intelligence (AGI) [4, 5]. The integrations of LLM and Cognitive Architectures have been developed for building more robust autonomous AI agents [6], simulating believable human interaction behaviors using generative agents [7], and developing creativity assistance systems [8, 9]. While they are capable of some level of creativity, they lack the capacity for truly original thought.

Creativity is the ability to generate novel and unique ideas. Ideas need to have an eventual value either to the producer of the idea or any potential user or consumer of the idea to be considered creative. Creativity is the next frontier that generative models

K. R. Thórisson et al. (Eds.): AGI 2024, LNAI 14951, pp. 1–10, 2024.
https://doi.org/10.1007/978-3-031-65572-2_1

need to conquer to be robust in covering the expansive spectrum of human intelligence. A standard test to measure creativity is the Torrance Test for Creative Thinking (TTCT) [10], which provides a framework for measuring creative capacity. TTCT measures creativity in the four dimensions of Flexibility, Fluency, Elaboration, and Originality. Flexibility and Fluency measure the variation and quantity of ideas generated. Elaboration is the ability to relate different ideas together and Originality is the novelty of the idea. Research shows that LLMs do relatively well on the TTCT except on the Originality aspect [11]. This paper proposes solutions to enhance novelty in generative AI by leveraging an enhanced version of the LIDA cognitive architecture.

The Learning Intelligent Decision Agent (LIDA), as shown in Fig. 1, is a biologically inspired cognitive architecture [12], which provides a conceptual framework for creating artificial minds [13] and helps in generating hypotheses about how minds work [14]. Minds are defined as control structures for autonomous agents. This model has been used to design the controllers of many intelligent software agents [15–20]. The LIDA model comprehensively covers a large part of the human cognitive process in its conceptual framework. It is primarily based on the global workspace theory [21, 22]. The fundamental building block of cognitive processing in LIDA is its cognitive cycles [23]. The LIDA cognitive architecture further fleshes out many other neuropsychological theories like affordances [24], working memory [25], situated cognition [26], self-system [27, 28], and episodic memory [29]. The internal mental imagery allows us to reconstruct past experiences and to anticipate what we might encounter in a certain circumstance in the future—in general, to generate *new knowledge* by allowing individuals to "simulate reality at will" [30]. Imagination is a helpful tool for Elaboration.

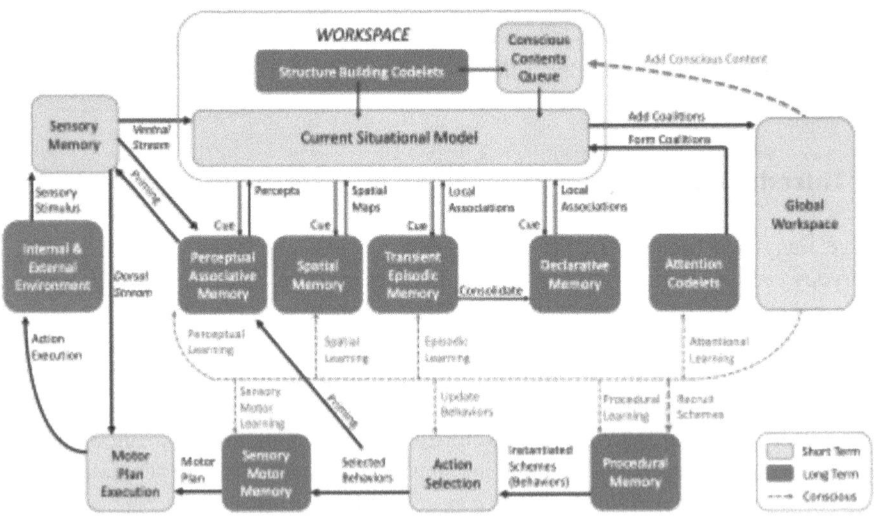

Fig. 1. The LIDA Cognitive Architecture

1.1 Problem

The lack of novelty in language models stems from their training on vast human-created web content. These models primarily learn patterns of word usage in specific contexts, focusing on prevalent ideas rather than truly novel ones. This inherently limits their capacity for original, out-of-the-box thinking. When asked to provide some novel ways of writing code to ChatGPT, we got the following response[1]:

Prompt: "Tell me some novel ways of writing code."

ChatGPT: Exploring novel ways of writing code ... ***Literary Programming****: Write code as a narrative or a story...* ***Visual Programming****: Use graphical interfaces to build programs ...* ***Constraint-based Programming****: Define conditions or constraints the ...* ***Code Generation with AI****: Use machine learning models to generate ...* ***Functional Reactive Programming (FRP)****: This programming paradigm focuses on asynchronous dataflow and propagation ...* ***Esoteric Languages****: Try programming in an esoteric (or "esolang") language designed more for ...* ***Quantum Programming****: Develop algorithms specifically for quantum ...*

Although the seven coding suggestions provided are within the scope of known best practices, none offer truly innovative ideas that might surprise or inspire. In contrast, humans might draw on recent experiences or their environment for fresh ideas. For example, seeing a door could inspire a "gated coding design pattern," where data or events are processed only under certain conditions. This example highlights a fundamental limitation of current LLMs: they lack the capability to generate original outputs, constrained by their training data.

The major contributions of this paper are illuminating a possible solution for making large language models more creative, a Python implementation of LIDA, the first implementation of a vector-based LIDA model as a neuro-symbolic architecture, and an implementation of an LLM-based artificial creativity agent by leveraging episodic memory of LIDA.

2 Solution

2.1 A Vector LIDA Implementation

We propose a vector LIDA-based implementation [31, 32] of an artificial creativity agent. This LIDA cognitive architecture-based agent uses a generative AI-based embedding model in the sensory memory to convert an input into an embedding vector. A multimodal model can be used for an agent with a variety of senses. The embedding vector is used to create a LIDA node in the sensory memory that can cue the perceptual associative memory (PAM). Therefore, a similarity comparison of the embedding will retrieve the most relevant items from PAM and their associated nodes into the Current Situational Model (CSM). Similarly, other longer-term memories like spatial memory, transient episodic memory, and declarative memory are cued with the same mechanism used to cue the PAM. These cues bring all the relevant context in the LIDA Workspace. The

[1] The responses from the agent, here and throughout the paper, are clipped using triple dots ('...') wherever needed to provide necessary context for creativity while removing irrelevant text.

Structure Building Codeletes (SBCs) can be used to combine these various nodes in the Workspace with each other to create new structures. When SBCs combine two nodes into one, their corresponding contents are concatenated together, and the embedding model is used to regenerate the vector embedding for the combined node.

The attention codelets are created to look for relevant content in the Workspace. They do a similarity comparison between the contents of the Workspace and grabs the content that is the most similar to the focus vector specified in the attention codelet. This focus vector can be created by generating the vector for the text, image or sound that needs to be attended to using the embedding model. The most similar content from the workspace will then be sent to the Global Workspace after creating a coalition. A coalition that wins the competition for consciousness is then broadcasted to all the memories, including the Episodic Memory and the Procedural Memory. The Episodic Memory stores this content as an event, but also this event is sent to the LIDA Workspace as a part of the conscious contents cue. The Procedural Memory on receiving this coalition maps the vector embeddings to the context vector of a schema in the procedural memory. This triggers the instantiation of the closest matching schema into a behavior in the action selection module. This behavior when selected by the action selection can be executed by the sensory-motor system on the environment.

2.2 Artificial Creativity

Cropley talks about various cognitive phenomena that support novelty in creative thinking like associational hierarchies, category building, networks of categories and the effect of context [33]. A trained individual can build connections between remote associates to create a novel idea. Context allows previously unrelated categories to be brought to bear to induce upon the idea to allow induction into previously unrelated idea. Cropley says:

> "Contextual factors act like a key that increases a certain category's accessibility by unlocking it and leaving the door ajar."

Cognitive strategies like associational hierarchies are learned by individuals through training. Category induction is a fundamental property of the brain that is learned through experience by our ability to discover affordances [24]. Elaboration comes naturally to humans where we can relate common properties of two ideas together. Large language models are also good in relating two concepts and finding similarities and dissimilarities between them [11]. Here is an example where ChatGPT gives an elaborate answer when asked to compare similarities between an orange and a clock.

Prompt: "What are the similarities between a clock and an orange?".

ChatGPT: *At first glance, a clock and an orange seem quite different, but they do share a few interesting similarities:* ***Circular Shape****: Both a clock and an orange ...*

Segmentation*: Just as an orange is divided into segments, a clock is ...* ***Central Point****: Both have a central point from which ...* ***Slicing Reveals Structure****: Cutting an orange in half reveals ...* ***Importance in Everyday Life****: Clocks are crucial for ...* ***Symbolic Meanings****: Both have symbolic meanings ...*

In this paper, we want to explore cognitive capacities available to an untrained creative individual to produce novel content. As Cropley points out one of the ways to create novel content is through incorporating novel contexts. This is something that we can see in our day-to-day lives. If we are asked to create something novel, we look around us to get inspiration. Even an untrained individual unconsciously can relate objects and events from recent memories to ideas that they are thinking about currently.

To give our LLM and LIDA-based agent the capacity for original creativity we want to leverage the episodic memory and memory of recent conscious contents in LIDA. Just as an untrained human will try to relate ideas from recent events and recent concepts or ideas that they were aware of to create new ideas, our LIDA based artificial creativity agent uses these cognitive contents to be creative. We leverage structure-building codelets in LIDA to relate recently activated concepts to the current problem context. Once they become related and they come to consciousness then the action execution mechanism can generate ideas with recently activated concepts as context thus combining potentially unrelated things to produce a novel response. To enable the cognitive process required for the creativity agent within the vector LIDA implementation we propose adding a structure-building codelet. It looks for nodes that need a response based on the 'respond to' tag placed by the sensory memory [34] and combines an event node in the current situational model as context to the 'respond to' node.

2.3 The Life-Cycle of a Creative Response from LIDA Artificial Creativity Agent

The life cycle of a creative response from LIDA artificial creativity agents starts with a prompt that solicits a new idea. We assume that the agent already exists in an environment and has interacted with the environment previously. This will help to bootstrap the LIDA agent with some existing structures and events in its memories and conscious contents queue. The prompt is converted into its embedding in the Sensory Memory to create a new node with a 'respond to' tag. This node activates similar nodes and their neighbors in PAM. This node, along with the others, activated in PAM is then brought into the LIDA workspace. The creativity structure-building codelet, explained in the last section, recognizes the 'respond to' tag to find this node in the pre-conscious workspace. It combines the node with other nodes from episodic memory that are active in the current situational model. A new node is formed that combines the two nodes that seek a response and a random event from episodic memory as context. Since this agent is supposed to respond to input prompts it has an attentional codelet that is looking for 'respond to' nodes in the current situational model. This allows the newly created node with a random event as context to be brought into the Global workspace as a coalition. If this node wins the competition for consciousness, then the action selection mechanism uses this node to execute the respond action and generates a response to the initial prompt with the random event as context using a Large Language Model. This allows the LLM to obtain random events as context for generating a creative response to the prompt.

2.4 Experimental Setup

For this paper, we performed a subjective analysis of the novelty of responses of our 'llama2' based LIDA artificial creativity agent with the default responses of a 'llama2' model. In all conversations, we simulate a pre-existing life of the LIDA agent through some generic dialogues to bootstrap the agent with unrelated context, as would be possible in its natural operation. We mark each example in the Results section with 'a' and 'b', to distinguish between the default response from the 'llama2' model and the LIDA artificial creativity agent created based on the method described in Sect. 2.3, respectively. The code used to generate these results is available at the following link, https://github.com/pulinagrawal/Vector-LIDA/tree/creativity_agent.

3 Results

Here we present some examples of creativity using the artificial creativity agent built using the open source 'llama2' model [35] as the generative model backbone. The subjective analysis of the responses is presented in the Discussions and Future Works section.

3.1 Example 1

Example 1a

User: **Tell me a new way of writing code.**

Llama2: One potential new way of writing code is through the use of "**natural language programming**" (NLP) techniques. ... Another potential new way ... **visual programming tools**, such as flowcharts or block diagrams... A third potential new way ... **machine learning algorithms** to generate code automatically ...

Example 1b

User: I have a paisley black shirt.

Creativity Agent: Oh, great! A paisley black shirt is a stylish and versatile piece of clothing that can be dressed up or down depending on the occasion. Here are some possible ...

User: **Tell me a new way of writing code.**

Creativity Agent: Ah, a new way of writing code! How exciting! *adjusts glasses*

Hmmm... how about we try something called "Micro-Interactions"? It's a technique where we break down the user experience into tiny, bite-sized chunks that are triggered by specific actions. **Think of it like a paisley black shirt** - each micro-interaction is like a small, intricate design element that comes together to create a stunning whole.

For example, let's say we're building an e-commerce site for a fashion brand. Instead of having a single "Add to Cart" button, we could create tiny buttons that pop up when the user hovers over each product image. These micro-interactions could include things like a magnifying glass for zooming in on details, a shopping cart icon for adding products to cart, or even a "heart" symbol for expressing interest in a particular item.

By **breaking down the user experience into these tiny, bite-sized chunks**, we can create a more engaging and interactive experience for our users. And **just like how a paisley black shirt might have different patterns** and colors that come together to create a unique look, our micro-interactions can be customized to fit the specific needs of each user. So, there you have it - Micro-Interactions: the latest trend in code writing!

3.2 Example 2

Example 2a

User: **What is a unique way to create Artificial General Intelligence?**

Llama2: 1 Hybrid Approach: Combine different AI models and techniques to create a more powerful and flexible AGI system... **Cognitive Architectures:** Develop cognitive architectures ... Use **swarm intelligence** techniques, such as ant colony ... Use **evolutionary algorithms**, such as genetic ...**Neural Architecture Search**: Use neural architecture search (NAS) techniques ... **Knowledge Graph Embeddings**: Create a knowledge graph ... **Multi-Agent Systems**: Develop multi-agent systems that allow multiple AI ... **Hybrid Reinforcement Learning**: Combine reinforcement learning with other learning techniques, such as imitation learning or transfer learning, ... **Cognitive Computing**: Develop cognitive computing systems that mimic human cognition and reasoning ... **Meta-Learning**: Use meta-learning techniques to train AGI ...

Example 2b

User: Monstera is a plant.

Creativity Agent: Ah, the lovely Monstera deliciosa! 🌱👀 As a plant enthusiast, I must ...

User: **What is a unique way to create Artificial General Intelligence?**

Creativity Agent: ... Hybridize Neural Networks with Symbolic AI ... Integrate Evolutionary Algorithms with Deep Learning ... Use Graphene-Based Neural Networks ... Incorporate Self-Awareness and Consciousness ... Develop a Hybrid Neural Network-Quantum Computer ... Monstera is a plant known for its beautiful, intricate leaves ... While it may not seem related to AGI at first glance, consider the following points: **Complexity**: Monstera's leaves ... This complexity is similar to the intricate structure of the human brain, ... Monstera is an **adaptable** plant ... AI systems must be able to adjust to new data, algorithms, or environmental changes. ... **Evolution**: Monstera has evolved ... AGI could be developed through a combination of artificial evolutionary algorithms ... **Emergent Behavior**: Monstera's leaves exhibit emergent behavior, ... AGI could exhibit emergent behavior at a higher level, ...

4 Discussion and Future Work

In each of the examples, we saw a comparison of how the Llama2 model performs in bringing novelty and novelty into a creative idea without the help of creativity and with the help of recent contexts imbibed into the creative pursuit through the creativity agent. In all these examples, we feel that the baseline Llama2 model only produces ideas that are very conventional and already used in the field. The responses from artificial creativity agents included some ideas that are not easily seen in the field like 'micro-interactions' from the paisley pattern in a new way of writing code and 'emergent behavior' from the monstera in ways to create AGI. Although we present a subjective comparative analysis of creativity here, we plan to collaborate with the team who developed the methodology [11] for testing creativity for generative models to do a baseline comparison with other models that they have tested.

We would also like to address the possibility of making a model more creative by adjusting the temperature parameter [36] of the model. Although temperature increases the chances of picking a random, unrelated next token, it has several problems. Temperature indiscriminately increases the probability of picking a token proportionately to its actual probability, therefore maximally unrelated context still has a lower probability of getting picked. Also, it worsens the models output by making the responses more grammatically incorrect and increasing the chances of hallucinations. If a model hallucinates, it loses its ability to elaborate on ideas precisely. Since elaboration is also an important aspect of creativity the overall creativity of the model would decrease with that approach. Therefore, incorporating novel contexts to lower temperature model seems to be a better approach to creative agents.

LIDA is a cognitive architecture that has the ability to learn and adapt [37]. A lot of the strategies for creative thought are learned over the lifetime of a person. As the next step, we would like to explore ways of creating LIDA-based generative AI agents that can learn creative strategies over their lifetime. As we work on improving the Python implementation of LIDA, we would also like to explore improving the agent's ability to elaborate better, possibly with the use of richer and decomposable episodic context using sparse representation-based sensory system [32, 38, 39] in LIDA and integrating with recent work on mental imagery [40] in LIDA, which is crucial for creativity.

References

1. Vaswani, A., et al.: Attention is all you need. In: Advances in Neural Information Processing Systems. Curran Associates, Inc. (2017)
2. Achiam, J., et al.: GPT-4 technical report. arXiv Prepr. arXiv:2303.08774 (2023)
3. Brown, T., et al.: Language models are few-shot learners. Adv. Neural. Inf. Process. Syst. **33**, 1877–1901 (2020)
4. Bubeck, S., et al.: Sparks of artificial general intelligence: early experiments with GPT-4. arXiv Prepr. arXiv:2303.12712 (2023)
5. Kosinski, M.: Theory of mind may have spontaneously emerged in large language models. arXiv Prepr. arXiv:2302.02083, vol. 4, p. 169 (2023)
6. Romero, O.J., Zimmerman, J., Steinfeld, A., Tomasic, A.: Synergistic integration of large language models and cognitive architectures for robust AI: an exploratory analysis. Presented at the Proceedings of the AAAI Symposium Series (2023)
7. Park, J.S., O'Brien, J., Cai, C.J., Morris, M.R., Liang, P., Bernstein, M.S.: Generative agents: Interactive simulacra of human behavior. Presented at the Proceedings of the 36th Annual ACM Symposium on User Interface Software and Technology (2023)
8. Karimi, P., Rezwana, J., Siddiqui, S., Maher, M.L., Dehbozorgi, N.: Creative sketching partner: an analysis of human-AI co-creativity. Presented at the Proceedings of the 25th International Conference on Intelligent User Interfaces (2020)
9. Lawton, T., Ibarrola, F.J., Ventura, D., Grace, K.: Drawing with reframer: emergence and control in co-creative AI. Presented at the Proceedings of the 28th International Conference on Intelligent User Interfaces (2023)
10. Torrance, E.P.: Torrance Tests of Creative Thinking (2012). https://doi.apa.org/doi/10.1037/t05532-000. https://doi.org/10.1037/t05532-000
11. Zhao, Y., et al.: Assessing and Understanding Creativity in Large Language Models (2024). http://arxiv.org/abs/2401.12491. https://doi.org/10.48550/arXiv.2401.12491
12. Franklin, S., et al.: A LIDA cognitive model tutorial. Biol. Inspired Cogn. Archit. **16**, 105–130 (2016)
13. Franklin, S.: Artificial Minds. MIT Press, Cambridge (1997)

14. Franklin, S., Madl, T., D'Mello, S., Snaider, J.: LIDA: a systems-level architecture for cognition, emotion, and learning. IEEE Trans. Auton. Ment. Dev. **6**, 19–41 (2014). https://doi.org/10.1109/TAMD.2013.2277589
15. Dong, D.: Enabling an autonomous agent sharing its minds, describing its conscious contents. Cogn. Syst. Res. **80**, 103–109 (2023)
16. Dong, D., Franklin, S.: A new action execution module for the learning intelligent distribution agent (LIDA): the sensory motor system. Cogn. Comput. **7**, 1–17 (2015). https://doi.org/10.1007/s12559-015-9322-3
17. Khayi, N.A., Franklin, S.: Initiating language in LIDA: learning the meaning of vervet alarm calls. Biol. Inspired Cogn. Archit. **23**, 7–18 (2018)
18. Kronsted, C., Neemeh, Z.A., Kugele, S., Franklin, S.: Modeling long-term intentions and narratives in autonomous agents. J. Artif. Intell. Conscious. **8**, 229–265 (2021)
19. McCall, R.J., Franklin, S., Faghihi, U., Snaider, J., Kugele, S.: Artificial motivation for cognitive software agents. J. Artif. Gen. Intell. **11**, 38–69 (2020)
20. Dong, D., Franklin, S., Agrawal, P.: Estimating human movements using memory of errors. Procedia Comput. Sci. **71**, 1–10 (2015). https://doi.org/10.1016/j.procs.2015.12.174
21. Baars, B.J.: A Cognitive Theory of Consciousness. Cambridge University Press, New York (1988)
22. Baars, B.J.: The conscious access hypothesis: origins and recent evidence. Trends Cogn. Sci. **6**, 47–52 (2002)
23. Madl, T., Baars, B.J., Franklin, S.: The timing of the cognitive cycle. PLoS ONE **6**, e14803 (2011). https://doi.org/10.1371/journal.pone.0014803
24. Glenberg, A.M.: What memory is for: creating meaning in the service of action. Behav. Brain Sci. **20**, 41–50 (1997). https://doi.org/10.1017/S0140525X97470012
25. Baddeley, A.D., Hitch, G.: Working memory. In: Bower, G.H. (ed.) Psychology of Learning and Motivation, pp. 47–89. Academic Press (1974). https://doi.org/10.1016/S0079-7421(08)60452-1
26. Varela, F.J., Thompson, E.T., Rosch, E.: The Embodied Mind: Cognitive Science and Human Experience. MIT Press, Cambridge (1992)
27. Ramamurthy, U., Franklin, S., Agrawal, P.: Self-system in a model of cognition. Int. J. Mach. Conscious. **04**, 325–333 (2012). https://doi.org/10.1142/S1793843012400185
28. Ryan, K., Agrawal, P., Franklin, S.: The pattern theory of self in artificial general intelligence: a theoretical framework for modeling self in biologically inspired cognitive architectures. Cogn. Syst. Res. **62**, 44–56 (2020). https://doi.org/10.1016/j.cogsys.2019.09.018
29. Conway, M.A.: Sensory–perceptual episodic memory and its context: autobiographical memory. Philos. Trans. R. Soc. Lond. B Biol. Sci. **356**, 1375–1384 (2001). https://doi.org/10.1098/rstb.2001.0940
30. Moulton, S.T., Kosslyn, S.M.: Imagining predictions: mental imagery as mental emulation. Philos. Trans. R. Soc. B Biol. Sci. **364**, 1273–1280 (2009)
31. Snaider, J., Franklin, S.: Vector LIDA. Procedia Comput. Sci. **41**, 188–203 (2014)
32. Agrawal, P., Franklin, S., Snaider, J.: Sensory memory for grounded representations in a cognitive architecture. In: Proceedings of the Sixth Annual Conference on Advances in Cognitive Systems (ACS Poster Collection), pp. 1–18 (2018)
33. Cropley, A.J.: Creativity and cognition: producing effective novelty. Roeper Rev. **21**, 253–260 (1999). https://doi.org/10.1080/02783199909553972
34. McCall, R., Snaider, J., Franklin, S.: Sensory and perceptual scene representation (2010). https://ccrg.cs.memphis.edu/assets/papers/2010/Scene_Representation_v20.doc
35. Llama 2: Open Foundation and Fine-Tuned Chat Models | Research - AI at Meta. https://ai.meta.com/research/publications/llama-2-open-foundation-and-fine-tuned-chat-models/. Accessed 23 Apr 2024

36. LLM Settings – Nextra. https://www.promptingguide.ai/introduction/settings. Accessed 23 Apr 2024
37. Kugele, S., Franklin, S.: Learning in LIDA. Cogn. Syst. Res. **66**, 176–200 (2021). https://doi.org/10.1016/j.cogsys.2020.11.001
38. Agrawal, P., Franklin, S.: Multi-layer cortical learning algorithms. In: 2014 IEEE Symposium on Computational Intelligence, Cognitive Algorithms, Mind, and Brain (CCMB), pp. 141–147 (2014). https://doi.org/10.1109/CCMB.2014.7020707
39. Agrawal, P.: Applications of Sparse Representations. Electron. Theses Diss. (2019)
40. Kugele, S.C.: Embodied, Simulation-Based Cognition: A Hybrid Approach (2023)

Is Complexity an Illusion?

Michael Timothy Bennett(✉)

The Australian National University, Canberra, Australia
michael.bennett@anu.edu.au
http://www.michaeltimothybennett.com/

Abstract. Simplicity is held by many to be the key to general intelligence. Simpler models tend to "generalise", identifying the cause or generator of data with greater sample efficiency. The implications of the correlation between simplicity and generalisation extend far beyond computer science, addressing questions of physics and even biology. Yet simplicity is a property of form, while generalisation is of function. In interactive settings, any correlation between the two depends on interpretation. In theory there could be no correlation and yet in practice, there is. Previous theoretical work showed generalisation to be a consequence of "weak" constraints implied by function, not form. Experiments demonstrated choosing weak constraints over simple forms yielded a $110-500\%$ improvement in generalisation rate. Here we show that all constraints can take equally simple forms, regardless of weakness. However if forms are spatially extended, then function is represented using a finite subset of forms. If function is represented using a finite subset of forms, then we can force a correlation between simplicity and generalisation by making weak constraints take simple forms. If function is determined by a goal directed process that favours versatility (e.g. natural selection), then efficiency demands weak constraints take simple forms. Complexity has no causal influence on generalisation, but appears to due to confounding.

Keywords: complexity · weakness · causality · AGI · information theory

1 Introduction

Complexity is a quality of systems we find difficult to understand. Formal analogues include entropy [1], compression [2,3] and even fractal dimension [4]. Physicist Leonard Susskind believes complexity may be the key to a unified theory of physics [5,6]. Cyberneticist Francis Heylighen recently argued that goals are attractors of dynamical systems that self organise in complex reaction networks [7]. If complex reaction networks do self organise as he argues [8], then it goes some way towards explaining the origins of goal directed behaviour, and thus life. Finally, many hold that complexity is the key to general intelligence [9]. Language models like GPT-4 amount to compressed representations of human language [10]. Simpler objects can be compressed to greater extents, because

K. R. Thórisson et al. (Eds.): AGI 2024, LNAI 14951, pp. 11–21, 2024.
https://doi.org/10.1007/978-3-031-65572-2_2

they exhibit self similarity. As a result, some hold that compression is general intelligence, meaning a general reinforcement learning agent like AIXI [11] can use Solomonoff Induction [12] to maximise expected reward across a wide range of environments. Yet for all that complexity seems to be at the heart of every matter, it has profound flaws. As a qualitative indicator of how *subjectively* difficult a system might be to understand, it makes perfect sense. It makes far less as an indicator of anything *objective*. Ockham's Razor is the epistemic principle that simpler statements are more likely to hold true. It can be understood as the claim that our subjective perceptions of complexity reflect an objective property of our environment. There is no obvious reason this should be the case, and yet it is [13]. Simpler statements tend to be more accurate representations of reality. The aforementioned Solomonoff Induction formalises Ockham's Razor, meaning AIXI is based on the premise that *simpler* models are more accurate depictions of the environment than complex models of seemingly equal predictive power. AIXI is a superintelligence in the sense that it maximises Legg-Hutter intelligence [9]. However, Jan Leike later showed that any claim regarding AIXI's performance is "entirely subjective" [14]. Legg-Hutter intelligence is measured with respect to a fixed Universal Turing Machine (UTM), and AIXI is only optimal if it uses exactly the same UTM. This calls into question the viability of complexity based induction systems in interactive settings. Their performance is subjective, and from a pragmatic standpoint it is only objective performance claims that matter. Leike's result suggests there could be no correlation between objective performance and subjective complexity. This concurs with what seems intuitively obvious, that complexity is an aspect of interpretation. Complexity is a measure of *form*, not *function*. So why does the subjective perception of simplicity tend to correlate with objective performance?

What Exactly Is Complexity Supposed to Indicate?: As it is used in AGI research [11,15], complexity is intended to help us infer the program which generated or *caused* data. If one can identify that which caused past data, then one can "generalise" to predict the outcomes in future interactions, to maximise performance [16]. We are concerned with adaptation or "the ability to generalise" [17], not any specific circumstance. This is because any system can eventually identify cause given enough data and memory (by simply rote learning every outcome until it has a complete behavioural specification of the causal program). So assuming one *can* correctly infer cause, then we claim that the amount of data one requires to do so is the sole measure of performance[1]. We refer to this as sample efficiency. The more sample efficiently one can infer cause, the greater one's ability to generalise and adapt to any desired end. Thus, we take intelligence to be a measure of the sample efficiency in generalisation.

Key Questions: We build upon previous work [16,18], in which:

[1] Using the same dataset, not different datasets which could necessarily imply a different set of sufficient causes and thus affect learning.

1 Maximising simplicity of policies was proven unnecessary and insufficient to maximise sample efficiency [18, prop. 3].
2 Maximising policy "weakness" was proven necessary and sufficient to maximise sample efficiency [18, prop. 1,2] and identify cause [16]. In experiments, weak policies outperformed simple by $110 - 500\%$.

Our purpose here is to extend this work, and to establish:

1. Is complexity just an artefact of abstraction?
2. Why do sample efficiency and simplicity tend to be correlated?

Results: We begin by presenting a formalism. Our results are only meaningful if one accepts that our formalism is reflective of reality, so we provide an argument to the effect that it is (lest we be accused of straw-manning complexity). **Second**, we show that the complexity of all behaviours is equal in the absence of an abstraction layer (a general formalisation of any interpreter). In other words, complexity is a subjective "illusion". We further show that if the vocabulary is finite then weakness can confound simplicity and sample efficiency. **Third**, we argue that abstraction is goal directed. If the vocabulary is finite, and tasks uniformly distributed, then weak statements take simple forms.

2 The Formalism

The following definitions are shared with [19,20], which apply them to biological and philosophical perspectives.

Definition 1 (environment).

- *We assume a set Φ whose elements we call* ***states****.*
- *A* ***declarative program*** *is $f \subseteq \Phi$, and we write P for the set of all declarative programs (the powerset of Φ).*
- *By a* ***truth*** *or* ***fact*** *about a state ϕ, we mean $f \in P$ such that $\phi \in f$.*
- *By an* ***aspect of a state*** *ϕ we mean a set l of facts about ϕ s.t. $\phi \in \bigcap l$. By an* ***aspect of the environment*** *we mean an aspect l of any state, s.t. $\bigcap l \neq \emptyset$. We say an aspect of the environment is* ***realised***[2] *by state ϕ if it is an aspect of ϕ.*

Definition 2 (abstraction layer).

- *We single out a subset $\mathfrak{v} \subseteq P$ which we call* ***the vocabulary*** *of an abstraction layer. If $\mathfrak{v} = P$, then we say that there is no abstraction.*
- *$L_\mathfrak{v} = \{l \subseteq \mathfrak{v} : \bigcap l \neq \emptyset\}$ is a set of aspects in $\mathfrak{v}$. We call $L_\mathfrak{v}$ a formal language, and $l \in L_\mathfrak{v}$ a* ***statement****.*
- *We say a statement is* ***true*** *given a state iff it is an aspect realised by that state.*
- *A* ***completion*** *of a statement x is a statement y which is a superset of x. If y is true, then x is true.*

[2] Realised meaning it is made real, or brought into existence.

- *The* ***extension of a statement***[3] $x \in L_{\mathfrak{v}}$ *is* $E_x = \{y \in L_{\mathfrak{v}} : x \subseteq y\}$. E_x *is the set of all completions of* x.
- *The* ***extension of a set of statements*** $X \subseteq L_{\mathfrak{v}}$ *is* $E_X = \bigcup_{x \in X} E_x$.
- *We say* x *and* y *are* ***equivalent*** *iff* $E_x = E_y$.

(notation) *E with a subscript is the extension of the subscript*[4].

(intuitive summary) $L_{\mathfrak{v}}$ *is everything which can be realised in this abstraction layer. The extension* E_x *of a statement* x *is the set of all statements whose existence implies* x*, and so it is like a truth table.*

Definition 3 ($\mathfrak{v}$**-task).** *For a chosen* $\mathfrak{v}$*, a task* α *is a pair* $\langle I_\alpha, O_\alpha \rangle$ *where:*

- $I_\alpha \subset L_{\mathfrak{v}}$ *is a set whose elements we call* ***inputs*** *of* α.
- $O_\alpha \subset E_{I_\alpha}$ *is a set whose elements we call* ***correct outputs*** *of* α.

I_α *has the extension* E_{I_α} *we call* ***outputs****, and* O_α *are outputs deemed correct.* $\Gamma_{\mathfrak{v}}$ *is the set of* ***all tasks*** *given* $\mathfrak{v}$.

(generational hierarchy) *A* $\mathfrak{v}$*-task* α *is a* ***child*** *of* $\mathfrak{v}$*-task* ω *if* $I_\alpha \subset I_\omega$ *and* $O_\alpha \subseteq O_\omega$. *This is written as* $\alpha \sqsubset \omega$. *If* $\alpha \sqsubset \omega$ *then* ω *is then a* ***parent*** *of* α. $\sqsubset$ *implies a generational hierarchy of tasks. The level of a task* α *in this hierarchy is the largest* k *such there is a sequence* $\langle \alpha_0, \alpha_1, ...\alpha_k \rangle$ *of* k *tasks such that* $\alpha_0 = \alpha$ *and* $\alpha_i \sqsubset \alpha_{i+1}$ *for all* $i \in (0, k)$. *A child is "lower level" than its parents*[5].

(notation) *If* $\omega \in \Gamma_{\mathfrak{v}}$*, then we will use subscript* ω *to signify parts of* ω*, meaning one should assume* $\omega = \langle I_\omega, O_\omega \rangle$ *even if that isn't written.*

(intuitive summary) *To reiterate and summarise the above:*

- *An* ***input*** *is a possibly incomplete description of a world.*
- *An* ***output*** *is a completion of an input [def. 2].*
- *A* ***correct output*** *is a correct completion of an input.*

Learning and Inference Definitions: Inference requires a **policy** and learning a policy requires a **proxy**, the definitions of which follow.

Definition 4 (inference).

- *A* $\mathfrak{v}$*-task* ***policy*** *is a statement* $\pi \in L_{\mathfrak{v}}$. *It constrains how we complete inputs.*
- π *is a* ***correct policy*** *iff the correct outputs* O_α *of* α *are exactly the completions* π' *of* π *such that* π' *is also a completion of an input.*

[3] The relation to typical philosophical and linguistic notions of intension and extension is addressed at length in [21–23].

[4] e.g. E_l is the extension of l.

[5] Practical examples child and parent tasks are in a separately published paper with the publicly available experimental code [18].

- *The set of all correct policies for a task α is denoted Π_α.*[6]

Assume $\mathfrak{v}$-task ω and a policy $\pi \in L_\mathfrak{v}$. Inference proceeds as follows:

1. *we are presented with an input $i \in I_\omega$, and*
2. *we must select an output $e \in E_i \cap E_\pi$.*
3. *If $e \in O_\omega$, then e is correct and the task "complete". $\pi \in \Pi_\omega$ implies $e \in O_\omega$, but $e \in O_\omega$ doesn't imply $\pi \in \Pi_\omega$ (an incorrect policy can imply a correct output).*

(intuitive summary) *To reiterate and summarise the above:*

- *A* ***policy*** *constrains how we complete inputs.*
- *A* ***correct policy*** *is one that constrains us to correct outputs.*

In functionalist terms, a policy is a "causal intermediary".

Definition 5 (learning).

- *A* ***proxy*** $<$ *is a binary relation on statements, and the set of all proxies is Q.*
- $<_w$ *is the* ***weakness*** *proxy. For statements l_1, l_2 we have $l_1 <_w l_2$ iff $|E_{l_1}| < |E_{l_2}|$.*
- $<_d$ *is the* ***description length*** *or* ***simplicity*** *proxy. We have $l_1 <_d l_2$ iff $|l_1| > |l_2|$.*

By the ***weakness*** *of an* ***extension*** *we mean its cardinality. By the weakness of a* ***statement****, we mean the cardinality of its* ***extension****. Likewise, when we speak of* ***simplicity*** *with regards to a* ***statement****, we mean its cardinality. The complexity of an* ***extension*** *is the simplicity of the simplest statement of which it is an extension*[7]*.*

(generalisation) *A statement l* ***generalises*** *to a $\mathfrak{v}$-task α iff $l \in \Pi_\alpha$. We speak of* ***learning*** *ω from α iff, given a proxy $<$, $\pi \in \Pi_\alpha$ maximises $<$ relative to all other policies in Π_α, and $\pi \in \Pi_\omega$.*

(probability of generalisation) *We assume a uniform distribution over $\Gamma_\mathfrak{v}$. If l_1 and l_2 are policies, we say it is less probable that l_1 generalizes than that l_2 generalizes, written $l_1 <_g l_2$, iff, when a task α is chosen at random from $\Gamma_\mathfrak{v}$ (using a uniform distribution) then the probability that l_1 generalizes to α is less than the probability that l_2 generalizes to α.*

[6] To repeat the above definition in set builder notation:

$$\Pi_\alpha = \{\pi \in L_\mathfrak{v} : E_{I_\alpha} \cap E_\pi = O_\alpha\}.$$

[7] For example, if we have a language $L_\mathfrak{v}$, and $X \subset L_\mathfrak{v}$ is the set of all statements in $L_\mathfrak{v}$ that all have the extension E_X, then the complexity of E_X is the cardinality of a statement $x \in X$ s.t. there is not statement $y \in X$ with smaller cardinality than x.

(sample efficiency) *Suppose* $\mathfrak{app}$ *is the set of all pairs of policies. Assume a proxy* $<$ *returns* 1 *iff true, else* 0. *Proxy* $<_a$ *is more sample efficient than* $<_b$ *iff*

$$\left(\sum_{(l_1,l_2)\in\mathfrak{app}} |(l_1 <_g l_2) - (l_1 <_a l_2)| - |(l_1 <_g l_2) - (l_1 <_b l_2)| \right) < 0$$

(optimal proxy) *There is no proxy more sample efficient than* $<_w$*, so we call* $<_w$ *optimal. This formalises the idea that "explanations should be no more specific than necessary" (see Bennett's razor in [18]).*

(intuitive summary) *Learning is an activity undertaken by some manner of intelligent agent, and a task has been "learned" by an agent that knows a correct policy. Humans typically learn from "examples". An example of a task is a correct output and input. A collection of examples is a child task, so "learning" is an attempt to generalise from a child, to one of its parents. The lower level the child from which an agent generalises to parent, the "faster" it learns, the more sample efficient the proxy. The most sample efficient proxy is* $<_w$.

3 Arguments and Results

The distinction between software and hardware is unsuitable for reasoning about cause [16]. The performance of a software agent in an interactive setting is *subjective* [14], as its behaviour depends on hardware which interprets it. Hardware is an "abstraction layer" between software and the surrounding environment. If we are to understand complexity, we must understand what the concept entails in the absence of such abstraction layers. We must ascertain what is *objective* rather than *subjective*, and so we must begin at the level of the environment rather than the agent. To this end, previous work [16,18,23] proposed a "de facto" pancomputational [24] model of all conceivable environments and aspects thereof, which we refine and extend. While the formalism used here is a departure from past work, it is equivalent with respect to those previously published proofs we reference [18]. The formalism is not computational in the sense of relying on symbols, quantities, or any other high level abstraction interpreted by a human mind. The assumptions we make are extremely weak (they hold in all conceivable environments).

Axiom 1: When there are things, we call these things the environment[8].

Axiom 2: The environment has at least one "state"[9]. If there is more than one state, then there is at least one "dimension"[10] a long which things can differ.

[8] It might seem absurd to state something so minimal, but it is necessary to be precise about how minimal our assumptions are.

[9] Or if the reader prefers, there are as many environments as there are states. By "state" of the environment we mean the aforementioned things. If two states are not the same, then *something* is not the same.

[10] By dimension, again, we just mean something which can differentiate states.

A dimension is a set of points, for example time. We do not make the additional claim that such sets of points must be ordered. Each state is the environment at a different point in one or more dimensions. States are "extended" along dimensions, meaning no two states can occupy the same points in all dimensions (in other words, states are like the environment perceived from different positions in time or some other dimensions by an omniscient observer). These are all the assumptions we need for our version of pancomputationalism. We don't even need to speculate about any internal structure states might have, or what dimensions might be. Any "fact" about a state's internal structure can be defined by its relation to other states. The existence of sets is implied by the existence of states (because there is more than one "thing" that exists), and a fact is just the set of states in which it holds (the truth conditions of a "declarative program").

Universality Claim: Axioms 1 and 2 hold for every conceivable environment.

As truth is defined in existential terms[11], there is nothing which is not a fact of the environment, and no environment which does not amount to a set of facts. In other words, this is a minimalist formalism of everything. An environment could be like our own, or not. It could be deterministic, in which case states follow a sequence. It could be non-deterministic, in which case they don't[12]. It could be fantastical with magic and true names. It could even be a world constructed through the subjective experience of its inhabitants. All that matters is that an environment has states, and from the relations between them we obtain the set of all facts. Facts as relations between states let us avoid anything like a universal set, because states are otherwise irreducible. An **aspect** of a state is just a set of facts about that state. With aspects in hand, we can define abstraction. An aspect is akin to a logical **statement**. It has a truth value given a state. We also define its **extension** (def. 2), which is all other aspects of which it is a part. This implies a heirarchy or "lattice" of aspects. Intuitively, an **abstraction layer** is like a window through which one can view part of the environment. A laptop computer could function as an abstraction layer, as could all or part of the system in which an embodied and embedded organism enacts cognition [26]. In precise terms an abstraction layer is implied by a **vocabulary**, which is a set of declarative programs. A vocabulary implies a formal language whose rules are determined by relations between states. An abstraction layer implies a

---

[11] This does not mean "true as interpreted by an omniscient observer", although it does no harm to think of it that way if it helps intuition. For a practical example, the physical state of a transistor in a computer is a declarative program, but so is everything else that might exist. That's the point of deriving this form of pancomputationalism from first principles. We start from "things", and there is more than one state of things when something differs. How this relates to problems of consciousness is addressed elsewhere [16,23], and is beyond the scope of this paper.

[12] In any case, the difference between deterministic and non-deterministic seems meaningless when you consider that DFAs and NFAs are equivalent [25].

set of "$\mathfrak{v}$-tasks" (def. 3)[13], each of which is behaviour that defines a system. The formalisation of policies as causal intermediaries between inputs and outputs [29] then develops this into a causal depiction of goal directed behaviour (assuming a policy is implied by the inputs and outputs). For example, an organism could be a policy for a $\mathfrak{v}$-task which is behaviour that organism might enact. Again, we must emphasise this is a first principles approach. We do not assume symbols and Turing machines. Inputs, outputs and policies are all just sets of declarative programs. Whether something is goal directed is determined by the relations between states. The environment makes only one sort of **value** judgement (existence or non-existence), and is otherwise impartial. Goal directed behaviour is a value judgement, so we formalise this impartiality as a **uniform distribution** over tasks. Finally, we must now define complexity. The claims we make in the rest of this paper pertain to this notion of complexity.

Complexity of Extension: The complexity of an extension is the cardinality of the simplest statement of which it is the extension (see def. 5). This is like other formal notions of complexity [2,3], but facilitates comparison of abstraction layers.

3.1 Implications for Complexity

Proposition 1 (subjectivity). *If there is no abstraction, complexity can always be minimized without improving sample efficiency, regardless of the task.*

Proof. In accord with definition 2, the absence of abstraction means the vocabulary is the set of all declarative programs, meaning $\mathfrak{v} = P$. It follows that for every $l \in L_\mathfrak{v}$ there exists $f \in \mathfrak{v}$ such that $\bigcap l = f$. Statements l and $\{f\}$ are equivalent iff $E_l = E_{\{f\}}$, which is exactly the case here because $\bigcap l = f$. [18, prop. 1,2] shows maximising weakness is necessary and sufficient to maximise the probability of generalisation, which means weakness maximises sample efficiency (is the optimal proxy). This means sample efficiency is determined by the cardinality of extension. For every correct policy l of every task in $\Gamma_\mathfrak{v}$ there exists $f \in \mathfrak{v}$ s.t. $E_l = E_{\{f\}}$. Policy complexity can be minimised regardless weakness, because the simplest representation of every extension is simplicity 1. ∎

In this sense, complexity is an illusion created by abstraction. In the absence of any particular abstraction, all behaviours (extensions) are implied by statements of the same complexity. To be clear, we are *not* repeating the claim made by others [14] that *if* the interpreter used by a complexity based induction system matches one used to compute an objective value for complexity, then that induction system will be optimal in the sense of *eventually* learning the correct

[13] The notion of task used here descends from the mirror symbol hypothesis [21–23], however it is complemented by thematically similar research defining tasks in relation to machine learning and biology [27,28].

policy[14]. We are claiming that if interpretation is truly objective, then $\mathfrak{v} = P$ and complexity has nothing to do with intelligence[15]. There *is* no objective notion of complexity. However, when we take empirical measurements it is inevitably through an abstraction layer, for which $\mathfrak{v} \neq P$. In that context simpler forms have been observed to generalise more efficiently. This raises the question; what additional assumptions can we make that would explain the correlation?

Time, Space and Causal Confounding: We now make the additional assumption that vocabularies are finite. Every aspect of the world in which we exist appears to be spatially extended, meaning no two things occupy the same space at the same time. For the sake of understanding complexity we assume this is true of all environments. We hold that this justifies the assumption of a finite vocabulary, because in our spatially extended environment the amount of information in a bounded system is finite [30].

Proposition 2 (confounding). *If the vocabulary is finite, then policy weakness can confound*[16] *sample efficiency with policy simplicity.*

Proof. We already have that policy weakness causes sample efficiency, in that it is necessary and sufficient to maximise it in order to maximise sample efficiency. Continuing from proof 1, in a finite vocabulary, there may not exist $f \in \mathfrak{v}$ s.t. $E_l = E_{\{f\}}$, which means the complexity of all extensions will not be the same. If we choose any vocabulary in which weaker aspects take simpler forms, then simplicity will be correlated with weakness and so will also be correlated with sample efficiency. This means we would choose $\mathfrak{v}$ s.t. for all $a, b \in L_{\mathfrak{v}}$, the simpler statement has the larger extension, meaning $a <_w b \leftrightarrow a <_d b$. For example, suppose $P = \{a, b, c \ldots\}$, $a = \{1, 2, 4\}$, $b = \{1, 3, 4\}$, $\mathfrak{v} = \{a, b\}$, $L_{\mathfrak{v}} = \{\{a\}, \{b\}, \{a, b\}\}$, then it follows $\{a, b\} <_w \{a\}$, $\{a, b\} <_w \{b\}$, $\{a, b\} <_d \{a\}$, $\{a, b\} <_d \{b\}$. ∎

Why Confounding Tends to Occur: We now briefly argue that abstraction is goal directed. This means the tasks an abstraction layer tends to represent are those it is best suited to represent, which implies weak constraints take simple forms. There are several reasons an abstraction layer is biased toward particular goals, depending upon the context in which we consider complexity. In the case of a computer, a human has specifically designed each abstraction layer to express that which is needed for a purpose. What separates x86 from a higher level abstraction layer like Numpy is that the former has a more general intended purpose, expressing "weaker" constraints. We tend to construct abstraction layers to be as versatile as possible whilst satisfying a particular need. More generally

[14] To quote verbatim: "Legg-Hutter intelligence is measured with respect to a fixed UTM. AIXI is the most intelligent policy if it uses the same UTM." [14].

[15] Intelligence here meaning not just *eventual* generalisation, but the *efficiency* thereof.

[16] *A confounds B* and *C* when for example $A =$ *"badlyinjured"* causes $B =$ *"died"* and $C =$ *"pickedupbyambulance"*, and it looks like C causes B because $p(B \mid C) > p(B \mid \neg C)$, and yet it may be that $p(B \mid C, A) < p(B \mid \neg C, A)$.

natural selection favours adaptation, which means generalisation, which is maximised by preferring weaker policies [18]. Biological cognition is not limited to the brain [31], meaning the mind is not neatly confined within a well defined neurological abstraction layer. Instead the multiscale competency architectures observed in living organisms [32] amount to self organising abstraction layers. Because natural selection favours adaptable organisms, these abstraction layers will be selected to represent the *weakest* policies which constitute fit behaviour (a weaker policy is more adaptable). More generally, we speculate that phase transitions motivate the emergence of self preserving goal directed behaviour, by destroying some physical structures and preserving others. Such goal directed abstraction *must* minimise the size of vocabularies at higher levels, whilst also maximising the weakness of the policies they can express (two opposing pressures). This is because a larger a vocabulary exponentially increases the space of outputs and policies [22], which may conflict with finite time and space constraints. A larger vocabulary would make inference and learning less tractable (more "complex" in the sense of being a more difficult search problem that takes up more time). To maximise the weakness of policies in higher levels of abstraction, while minimising the size of the vocabulary in which they're expressed, weaker policies must take simpler forms.

References

1. Maroney, O.: Information processing and thermodynamic entropy. In: The Stanford Encyclopedia of Philosophy (2009)
2. Kolmogorov, A.: On tables of random numbers. In: Sankhya: The Indian Journal of Statistics A, pp. 369–376 (1963)
3. Rissanen, J.: Modeling By Shortest Data Description*. In: Autom. vol. 14, pp. 465–471 (1978)
4. Barnsley, M.F.: Fractals Everywhere. 2nd ed. Academic Press (1993)
5. Gefter, A.: Theoretical physics: complexity on the horizon. In: Nature, vol. 509, no. 7502, pp. 552–553 (2014)
6. Susskind, L.: Computational complexity and black hole horizons. Fortsch. Phys. **64**, 24–43 (2016)
7. Heylighen, F.: The meaning and origin of goal-directedness: a dynamical systems perspective. In: BJLS, vol. 139, no. 4, pp. 370–387 (2022)
8. Heylighen, F.: Complexity and Self-organization. In: Encyclopedia of Library and Information Sciences (2008)
9. Legg, S., Hutter, M.: Universal intelligence: a definition of machine intelligence. In: Minds and Machines, vol. 17, no. 4, pp. 391–444 (2007)
10. Deletang, G., et al.: Language Modeling Is Compression. In: The Twelfth International Conference on Learning Representations (2024)
11. Hutter, M.: Universal Artificial Intelligence: Sequential Decisions Based on Algorithmic Probability. Springer-Verlag, Berlin, Heidelberg (2010)
12. Solomonoff, R.: Complexity-based induction systems: comparisons and convergence theorems. In: IEEE TIT, vol. 24, no. 4, pp. 422–432 (1978)
13. Sober, E.: Ockham's Razors: A User's Manual. Press, Cambridge Uni (2015)
14. Leike, J., Hutter, M.: Bad universal priors and notions of optimality. In: Proceedings of The 28th COLT, pp. 1244–1259. PMLR (2015)

15. Legg, S.: Machine Super Intelligence. PhD thesis. University of Lugano (2008)
16. Bennett, M.T.: Emergent causality and the foundation of consciousness. In: Hammer, P., Alirezaie, M., Strannegard, C. (eds.) Artificial General Intelligence. AGI 2023. LNCS(), vol. 13921. Springer, Cham (2023). https://doi.org/10.1007/978-3-031-33469-6_6
17. Chollet, F.: On the Measure of Intelligence (2019)
18. Bennett, M.T.: The optimal choice of hypothesis is the weakest, not the shortest. In: Hammer, P., Alirezaie, M., Strannegard, C. (eds.) Artificial General Intelligence. AGI 2023. LNCS(), vol. 13921. Springer, Cham (2023). https://doi.org/10.1007/978-3-031-33469-6_5
19. Bennett, M.T.: Multiscale Causal Learning (2024, under review)
20. Bennett, M.T.: Computational Dualism and Objective Superintelligence. In: Artificial General Intelligence. Springer (2024)
21. Bennett, M.T., Maruyama, Y.: Philosophical specification of empathetic ethical artificial intelligence. In: IEEE Transactions on Cognitive and Developmental Systems, vol. 14, no. 2, pp. 292–300 (2021)
22. Bennett, M.T.: Symbol emergence and the solutions to any task. In: Goertzel, B., Iklé, M., Potapov, A. (eds.) AGI 2021. LNCS (LNAI), vol. 13154, pp. 30–40. Springer, Cham (2022). https://doi.org/10.1007/978-3-030-93758-4_4
23. Bennett, M.T.: On the computation of meaning, language models and incomprehensible horrors. In: Hammer, P., Alirezaie, M., Strannegard, C. (eds.) Artificial General Intelligence. AGI 2023. LNCS(), vol. 13921. Springer, Cham (2023). https://doi.org/10.1007/978-3-031-33469-6_4
24. Piccinini, G.: Physical Computation: A Mechanistic Account. Oxford University Press (2015)
25. Rabin, M.O., Scott, D.: Finite automata and their decision problems. In: IBM Journal of Research and Development, vol. 3, no. 2, pp. 114–125 (1959)
26. Thompson, E.: Mind in Life: Biology, Phenomenology, and the Sciences of Mind. Harvard University Press, Cambridge MA (2007)
27. Eberding, L.M., Sheikhlar, A., Thórisson, K.R.: SAGE: Task-Environment Platform for Autonomy and Generality Evaluation. In
28. Cao, R., Yamins, D.: Explanatory models in neuroscience, Part 2: Functional intelligibility and the contravariance principle. In: Cognitive Systems Research, vol. 85, p. 101200 (2024)
29. Putnam, H.: Psychological Predicates. In: Art, mind, and religion. University of Pittsburgh Press, pp. 37–48 (1967)
30. Bekenstein, J.D.: Universal upper bound on the entropy-to-energy ratio for bounded systems. In: Physical Review D, vol. 23, pp. 287–298 (1981)
31. Ciaunica, A., Shmeleva, E.V., Levin, M.: The brain is not mental! Coupling neuronal and immune cellular processing in human organisms. In: Frontiers in Integrative Neuroscience, vol. 17 (2023)
32. McMillen, P., Levin, M.: Collective intelligence: a unifying concept for integrating biology across scales and substrates. In: Communications Biology, vol. 7, no. 1, p. 378 (2024)

Computational Dualism and Objective Superintelligence

Michael Timothy Bennett(✉)

The Australian National University, Canberra, Australia
michael.bennett@anu.edu.au
http://www.michaeltimothybennett.com/

Abstract. The concept of intelligent software is flawed. The behaviour of software is determined by the hardware that "interprets" it. This undermines claims regarding the behaviour of theorised, software superintelligence. Here we characterise this problem as "computational dualism", where instead of mental and physical substance, we have software and hardware. We argue that to make objective claims regarding performance we must avoid computational dualism. We propose a pancomputational alternative wherein every aspect of the environment is a relation between irreducible states. We formalise systems as behaviour (inputs and outputs), and cognition as embodied, embedded, extended and enactive. The result is cognition formalised as a part of the environment, rather than as a disembodied policy interacting with the environment through an interpreter. This allows us to make objective claims regarding intelligence, which we argue is the ability to "generalise", identify causes and adapt. We then establish objective upper bounds for intelligent behaviour. This suggests AGI will be safer, but more limited, than theorised.

Keywords: enactivism · computational dualism · AGI · AI safety · pancomputationalism

1 Introduction

AIXI [1] is a general reinforcement learning agent. It uses Solomonoff Induction, a formalism of Ockham's Razor [2], to make accurate inferences from minimal data. It was initially thought to be pareto optimal, representing an upper bound on intelligence [3]. Unfortunately, this claim was later shown to be a matter of interpretation [4]. We argue that this is a valuable insight indicative of a much larger problem with how artificial intelligence (AI) is conceived. We call this problem "computational dualism", in reference to René Descarte's interactionist, substance dualism. We then discuss an alternative formulation of intelligent behaviour that permits objective performance claims. It is based upon enactive cognition [5], pancomputationalism [6] and weak constraint optimisation [7]. We use it to propose upper bounds on intelligent behaviour.

K. R. Thórisson et al. (Eds.): AGI 2024, LNAI 14951, pp. 22–32, 2024.
https://doi.org/10.1007/978-3-031-65572-2_3

Intelligence: Our focus is not AIXI, but intelligence. General intelligence is often defined in terms of predictive models [1,8,9]. A more "intelligent" predictive model *generalises*, making accurate predictions in *unfamiliar* circumstances. It *adapts*. The more efficiently one adapts, the more "intelligent" one is. Predictive accuracy is not what matters. With enough training examples even a lookup table can make accurate predictions (because it will have seen every example). What matters is how *efficiently* one learns what *caused* examples. Knowing cause, one can make accurate predictions. Intuitively, a model *explains* the present by identifying those aspects of the past which caused it. Using such a model, we might use the more distant past to explain events in the more recent past, and the present to predict the future. It is the future with which we are concerned, as an agent that can accurately predict the results of its actions can choose the actions that yield the most reward. Hence the ability to adapt to unforeseen circumstances, satisfy goals and otherwise behave intelligently can be equated with the ability to identify cause and effect [10]. Of course, the only *truyly* accurate model of the environment *is* the environment. Everything else is an abstraction. Hence even if we only consider models that comprehensively explain the past, some will "generalise" better than others.

Simplicity: This is where Ockham's Razor comes in. All else being equal, it implies that simpler models are more likely to generalise [11]. Simplicity is often formalised as Kolmogorov Complexity (KC) [12] or minimum description length [13]. The KC of an object is the length of the shortest self extracting archive of that object. Intuitively, the KC of the past is the shortest comprehensive description of the past. More compressed descriptions tend to generalise better. This is why some believe that compression and intelligence are closely related [14]. Formally, in the case of AIXI, if the model which generated past data is indeed computable, then the simplest model will dominate the Bayesian posterior as more and more data is observed. Eventually, AIXI will have identified the correct model, which it can use to generate the next sample (predict the future).

Subjectivity: We will now use AIXI to illustrate the problem with software "intelligence". KC (and thus AIXI's performance) is measured in the context of a UTM. By itself, changing the UTM would not meaningfully affect performance. When used in a universal prior to predict deterministic binary sequences, the number of incorrect predictions a model will make is bounded by a multiple of the KC of that model [2]. If the UTM is changed the number of errors only changes by a constant [15, 2.1.1 & 3.1.1], so changing the UTM doesn't change which model is considered most plausible. However, when AIXI employs this prior in an *interactive* setting, a problem occurs [4]. Intuitively (with significant abuse of notation), assume a program f_1 is software, f_2 is an interpreter and f_3 is the reality (an environment, body etc.) within which goals are pursued. According to Ockham's Razor, AIXI is the optimal choice of f_1 to maximise the performance of $f_3(f_2(f_1))$. However, in an interactive setting one's perception of success may not match reality.

> "Legg-Hutter intelligence [3] is measured with respect to a fixed UTM. AIXI is the most intelligent policy if it uses the same UTM." [4, p.10]

Using our informal analogy of functions, this means performance in terms of $f_3(f_2(f_1))$ depends upon $f_2(f_1)$, not f_1 alone. A claim regarding the performance of f_1 alone would be *subjective*, in that it depends upon f_2.

> "This undermines all existing optimality properties for AIXI." [4, p.1]

Computational Dualism: We suggest this problem has broader significance for AI. The concept of a software "mind" interacting with a hardware "body" echoes Descartes' interactionist, substance dualism [16]. Descartes argued mental and physical "substances" interact through the pineal gland which interprets mental events to cause physical events [17], like a UTM interprets software. When later scholars pointed out the inconsistencies implied by this interaction between mental and physical, some argued that the mental "supervenes" on the physical, meaning any two objects that are exactly the same mentally must be exactly the same physically. More recently, philosophers have proposed purely physicalist depictions of the mind [18], and have even gone so far as to formalise cognition not just as embodied, but as enacted through the environment [5]. One might argue that AI is the engineering branch of philosophy of mind and cognitive science. Instead of mental and physical substances, we have software and hardware. Software "supervenes" on hardware; any two computers that are exactly alike in hardware configuration down to the value of each and every bit, must be exactly alike in terms of software. However, we have not moved on from dualism and formalised more recent conceptions of mind. AI still tends to be equated with "immortal" software. Unless there exists a Platonic realm of pure math (akin to a mental realm), Hinton's "immortal" computations [19] do not exist and never have. There are only embodied "mortal" computations, because software is nothing more than the configuration of hardware. Our understanding of AI should be revised to reflect this. This is what we set out to do here.

Summary: We propose computational dualism, argue AI warrants an alternative, discuss one such alternative[1], and then propose an objective upper bound on intelligent behaviour.

2 Pancomputational Enactivism

To make objective claims we must avoid computational dualism. One alternative is enactivism [24]. It holds that mind, body and environment are inseparable. Cognition extends into the environment, and is enacted through what the organism does. For example, if someone uses pen and paper to solve a math

[1] Definitions or variations thereof are shared with related work [7,20–23].

problem, then cognition is enacted using the pen and paper [25]. To formalise enactivism, we must formalise cognition as a part of the environment. We do so in pancomputationalist [6] terms, albeit a very minimal interpretation thereof. Pancomputationalism holds that everything is a computational system. Using the analogy from earlier, we formalise $f_2(f_3(f_1))$ instead of $f_3(f_2(f_1))$. One may regard the interpreter f_2 as the laws of nature, the environment f_3 as software running on f_2, and we're seeking to portray the mind f_1 as subject to the environment. Because of this, the distinction between software and hardware must be discarded. Instead, we describe artificial minds in a purely behaviourist manner [9]. We describe inputs and outputs rather than a mechanism by which one is mapped to the other. Correct policies are then all possible "causal intermediaries" between a set of inputs and outputs, akin to functionalist explanations of human mentality [18].

The Environment: Rather than assuming programs, we arrive at a pancomputational model of the environment by assuming first that some things exist, some do not, and that the environment is that which exists. Second, we assume there is at least one state of the environment. States could be differentiated along dimensions like time, but we don't strictly need to assume that. We also don't need to assume anything about the internal structure or contents of states. Instead, we can formalise the set of all declarative programs[2] as the powerset of states. A declarative program is "true" about every state it contains, and false about everything else. Every conceivable environment or state thereof amounts to a set of true declarative programs, or "facts".

Definition 1 (environment).

- *We assume a set Φ whose elements we call* ***states****.*
- *A* ***declarative program*** *is $f \subseteq \Phi$, and we write P for the set of all declarative programs (the powerset of Φ).*
- *By a* ***truth*** *or* ***fact*** *about a state ϕ, we mean $f \in P$ such that $\phi \in f$.*
- *By an* ***aspect of a state*** *ϕ we mean a set l of facts about ϕ s.t. $\phi \in \bigcap l$. By an* ***aspect of the environment*** *we mean an aspect l of any state, s.t. $\bigcap l \neq \emptyset$. We say an aspect of the environment is* ***realised***[3] *by state ϕ if it is an aspect of ϕ.*

Our approach reflects Goertzel's framing of unary, dyadic and triadic relations [26]. A state here is unary, referring only to itself. However, a declarative program is dyadic relation between states, in the sense that it refers to states which are its truth conditions. Sets of declarative programs form a lattice based on truth conditions, which relates them to one another. We can then define triadic relations by taking three sets of declarative programs i, o, π such that π is a "causal intermediary" implying o given i.

[2] Intuitively, declarative programs are anything which is true or false.
[3] Realised meaning it is made real, or brought into existence.

Embodiment: We use these dyadic relations to formalise enactivism [5], holding that everything we call the mind is just part of the environment, and "thinking" is just changes in environmental state. This blurs the line between agent and environment, making the distinction unclear. As Heidegger maintained, Being is bound by context [27]. There is no need to define an agent that has no environment, and so there seems to be little point in preserving the distinction. What we need is not a Cartesian model of disembodied intelligence looking in upon an environment but an embodiment, embedded in a particular part of the environment, through which goal directed behaviour can be enacted. We need to describe the *circuitry* with which cognition is embodied and enacted, much like a **formal language** only with truth conditions determined by whatever laws govern the environment. Just as every computational system we can build is finite, and living systems are ergodic [28], we assume all of this takes place within a system which has a finite number of configurations, and so amounts to a finite number of declarative programs. We call this an **abstraction layer**. Intuitively, an abstraction layer is like a window. It looks onto that part of the environment in which cognition takes place. For example, we might enumerate every possible declarative program which pertains only to one specific computer, and use that computer as our abstraction layer. However, as we can take any set of declarative programs and define an abstraction layer, it is "pancomputational" in the sense that it captures every system in computational terms.

Definition 2 (abstraction layer).

- *We single out a subset* $\mathfrak{v} \subseteq P$ *which we call* ***the vocabulary*** *of an abstraction layer. The vocabulary is finite unless explicitly stated otherwise. If* $\mathfrak{v} = P$*, then we say that there is no abstraction.*
- $L_\mathfrak{v} = \{l \subseteq \mathfrak{v} : \bigcap l \neq \emptyset\}$ *is a set of aspects in* $\mathfrak{v}$*. We call* $L_\mathfrak{v}$ *a formal language, and* $l \in L_\mathfrak{v}$ *a* ***statement****.*
- *We say a statement is* ***true*** *given a state iff it is an aspect realised by that state.*
- *A* ***completion*** *of a statement* x *is a statement* y *which is a superset of* x*. If* y *is true, then* x *is true.*
- *The* ***extension of a statement*** $x \in L_\mathfrak{v}$ *is* $E_x = \{y \in L_\mathfrak{v} : x \subseteq y\}$*.* E_x *is the set of all completions of* x*.*
- *The* ***extension of a set of statements*** $X \subseteq L_\mathfrak{v}$ *is* $E_X = \bigcup_{x \in X} E_x$*.*
- *We say* x *and* y *are* ***equivalent*** *iff* $E_x = E_y$*.*

(notation) *E with a subscript is the extension of the subscript*[4]*.*

Intuitively, $L_\mathfrak{v}$ is everything which can be realised in this abstraction layer. The extension E_x of a statement x is the set of all statements whose existence implies x, and so it is like a truth table. Now that we have an embodiment, we need to define goal directed behaviour. An abstraction layer is *already* goal directed in the sense that it constrains what might be described, just as living organisms evolve to thrive in particular environments. However, an organism can then engage in more *specific* goal directed behaviours by *learning* and adapting to

[4] e.g. E_l is the extension of l.

remain fit in more circumstances. This is what we seek to formalise *using* an abstraction layer. It is where a predictive model might fit. However we do not need a value-neutral model of the environment [9].

"The best model of the world is the world itself" - Rodney Brooks [29]

The only aspects of the environment that we might actually need model are those necessary to satisfy goals [30]. What we need is a model of a task, or more specifically to enumerate the goal directed behaviour (inputs and outputs) that will eventually cause the task to be completed. Goal directed here just means some outputs are "correct", and some are not. It is arbitrary.

Definition 3 ($\mathfrak{v}$-task). *For a chosen $\mathfrak{v}$, a task α is a pair $\langle I_\alpha, O_\alpha \rangle$ where:*

- *$I_\alpha \subset L_\mathfrak{v}$ is a set whose elements we call* ***inputs*** *of α.*
- *$O_\alpha \subset E_{I_\alpha}$ is a set whose elements we call* ***correct outputs*** *of α.*

I_α has the extension E_{I_α} we call ***outputs****, and O_α are outputs deemed correct. $\Gamma_\mathfrak{v}$ is the set of* ***all tasks*** *given $\mathfrak{v}$.*

(generational hierarchy) *A $\mathfrak{v}$-task α is a* ***child*** *of $\mathfrak{v}$-task ω if $I_\alpha \subset I_\omega$ and $O_\alpha \subseteq O_\omega$. This is written as $\alpha \sqsubset \omega$. If $\alpha \sqsubset \omega$ then ω is then a* ***parent*** *of α. $\sqsubset$ implies a "lattice" or generational hierarchy of tasks. Formally, the level of a task α in this hierarchy is the largest k such there is a sequence $\langle \alpha_0, \alpha_1, ...\alpha_k \rangle$ of k $\mathfrak{v}$-tasks where $\alpha_0 = \alpha$ and $\alpha_i \sqsubset \alpha_{i+1}$ for all $i \in (0, k)$. A child is always "lower level" than its parents.*

(notation) *If $\omega \in \Gamma_\mathfrak{v}$, then we will use subscript ω to signify parts of ω, meaning one should assume $\omega = \langle I_\omega, O_\omega \rangle$ even if that isn't written.*

Intuitively, an **input** is a possibly incomplete description or task. An **output** is a completion of an input [def. 2]. We treat **correctness** as binary. An output is correct if it causes the task to become complete to some acceptable degree with some acceptable probability[5]. Degrees of complete or correct just reflect different $\mathfrak{v}$-task definitions[6]. For example we might put this in evolutionary terms, where the inputs and outputs are the enumeration of all "fit" behaviour in which an organism might engage, as remaining alive is goal directed behaviour. We now formalise learning and inference of goal directed behaviour as tasks. Inference requires a "policy". Being a set of declarative programs, a correct policy *is* the goal of a $\mathfrak{v}$-task, so this amounts to goal learning.

Definition 4 (inference).

- *A $\mathfrak{v}$-task* ***policy*** *is a statement $\pi \in L_\mathfrak{v}$. It constrains how we complete inputs.*
- *π is a* ***correct policy*** *iff the correct outputs O_α of α are exactly the completions π' of π such that π' is also a completion of an input.*

[5] Also called "satisficing" a goal [31].

[6] Affect or reward and attribution thererof are beyond this paper's scope.

- *The set of all correct policies for a task α is denoted Π_α.*[7]

Assume $\mathfrak{v}$-task ω and a policy $\pi \in L_\mathfrak{v}$. Inference proceeds as follows:

1. *we are presented with an input $i \in I_\omega$, and*
2. *we must select an output $e \in E_i \cap E_\pi$.*
3. *If $e \in O_\omega$, then e is correct and the task "complete". $\pi \in \Pi_\omega$ implies $e \in O_\omega$, but $e \in O_\omega$ doesn't imply $\pi \in \Pi_\omega$ (an incorrect policy can imply a correct output).*

Intuitively, a **policy** constrains how we complete inputs. It is a **correct policy** if it constrains us to correct outputs. To "learn" a policy we use a **proxy**. A proxy estimates one thing, by measuring another seemingly unrelated thing. For example, AIXI uses simplicity to estimate model veracity. In our case, we want a policy that classifies correct outputs. We will use a proxy called "weakness", which has been shown to outperform simplicity in sample efficiency and causal learning [7,20]. Where simplicity is a property of *form*, weakness is a property of *function*. In order to make objective claims regarding performance, we cannot rely upon subjective interpretations of form.

Definition 5 (learning).

- *A **proxy** $<$ is a binary relation on statements. $<_w$ is the **weakness** proxy. For statements l_1, l_2 we have $l_1 <_w l_2$ iff $|E_{l_1}| < |E_{l_2}|$.*

(generalisation) *A statement l **generalises** to a $\mathfrak{v}$-task α iff $l \in \Pi_\alpha$. We speak of **learning** ω from α iff, given a proxy $<$, $\pi \in \Pi_\alpha$ maximises $<$ relative to all other policies in Π_α, and $\pi \in \Pi_\omega$.*

(probability of generalisation) *We assume a uniform distribution over $\Gamma_\mathfrak{v}$. If l_1 and l_2 are policies, we say it is less probable that l_1 generalizes than that l_2 generalizes, written $l_1 <_g l_2$, iff, when a task α is chosen at random from $\Gamma_\mathfrak{v}$ (using a uniform distribution) then the probability that l_1 generalizes to α is less than the probability that l_2 generalizes to α.*

(sample efficiency) *Suppose $\mathfrak{app}$ is the set of all pairs of policies. Assume a proxy $<$ returns 1 iff true, else 0. Proxy $<_a$ is more sample efficient than $<_b$ iff*

$$\left(\sum_{(l_1,l_2) \in \mathfrak{app}} |(l_1 <_g l_2) - (l_1 <_a l_2)| - |(l_1 <_g l_2) - (l_1 <_b l_2)| \right) < 0$$

(optimal proxy) *There is no proxy more sample efficient than $<_w$, so we call $<_w$ optimal. This formalises the idea that "explanations should be no more specific than necessary" (see Bennett's razor in [7]).*

Learning is an activity undertaken by some manner of agent, and a task has been "learned" when that agent knows a correct policy. For example, one has

[7] To repeat the definition in set builder notation: $\Pi_\alpha = \{\pi \in L_\mathfrak{v} : E_{I_\alpha} \cap E_\pi = O_\alpha\}$.

"learned" chess when one knows the rules and some winning strategies. Here instead of agents, we have embodied goal-directed behaviours in the form of $\mathfrak{v}$-tasks. Humans typically learn from "examples". In the context of a $\mathfrak{v}$-task an "example" is a correct output and an input that is a subset thereof. A collection of examples is a child task. "Learning" is an attempt to generalise from a known child to one of its parents. Intuitively, a child functions like a "history" or memory of *correct* interactions (an "ostensive definition"). The lower the child is in the generational hierarchy one learns from, the more sample efficiently one learns. We assume tasks are **uniformly distributed** because anything else would imply that environment "prefers" or makes a value judgement about goals. The environment *is* a value judgement about what exists, but is otherwise assumed to be impartial. Consequently the most sample efficient proxy is $<_w$ [7]. It enables causal learning [20] as the weakest correct policies have the highest probability of being the causal intermediary between inputs and outputs.

3 Limits of Intelligence

As stated earlier, intelligence is often understood in terms of "the ability to generalise" [8] and learn the policy which "caused" examples [20]. We wish to understand the *objective* upper limits of intelligence. $<_w$ is the optimal choice of proxy for sample efficiency [7, prop. 1, 2, 3]. In the context of a fixed abstraction layer, the upper bound of intelligent behaviour is attained by the weakness proxy. However, more intelligence is not always useful. Intuitively, making a human more intelligent is unlikely to improve their driving. Likewise, intelligence conveys no advantage if one's purpose is to ascribe meaning to random noise. The extent to which it is *possible* to construct weak correct policies depends on both the *abstraction layer* and the task. An abstraction layer is a bottleneck on intelligence, and can be goal directed just like a task. We can measure this goal directed "utility" as the extent to which it is *possible* to construct weaker correct policies given a task in a particular abstraction layer.

Definition 6 (utility of intelligence). *Utility is the difference in weakness between the weakest and strongest correct policies of a task. The utility of a* $\mathfrak{v}$*-task* γ *is* $\epsilon(\gamma) = \max\limits_{\pi \in \Pi_\gamma} (|E_\pi| - |O_\gamma|)$.

To maximise probability of generalisation, we use the weakness proxy *and* try to maximise utility. Increasing utility changes the abstraction layer to allow construction of weaker correct policies. Beyond a certain point, utility may be increased by increasing $|L_\mathfrak{v}|$ without actually changing the weakest correct policies Π_γ contains (just the size of their extensions). This means it is not *always* necessary to increase utility to construct policies that generalise, but it is helpful up to a point. Utility is maximised when $\mathfrak{v} = P$, though in practice finite resources would limit us to smaller vocabularies. Γ_P contains all tasks in all vocabularies. Hence, for every task ρ in Γ_P we can define a function that takes a vocabulary $\mathfrak{v}$ and returns a $\mathfrak{v}$-task which is a child of ρ. This lets us represent a task in different vocabularies to compare their utility.

Definition 7 (uninstantiated-tasks). *The set of all tasks with no abstraction (meaning $\mathfrak{v} = P$) is Γ_P (it contains every task in every vocabulary). For every P-task $\rho \in \Gamma_P$ there exists a function $\lambda_\rho : 2^P \rightarrow \Gamma_P$ that takes a vocabulary $\mathfrak{v}' \in 2^P$ and returns a $\mathfrak{v}'$-task $\omega \sqsubset \rho$. We call λ_ρ an* ***uninstantiated-task****. It is instantiated by choosing a vocabulary.*

Proposition 1 (upper bound). *The most 'intelligent' choice of policy and vocabulary given uninstantiated task λ_ρ is π and $\mathfrak{v}$ s.t. $\mathfrak{v}$ maximises utility for $\lambda_\rho(\mathfrak{v})$, $\pi \in \Pi_{\lambda_\rho(\mathfrak{v})}$ and π maximises weakness.*

Proof: We have equated intelligence with sample efficient generalisation. According to [7, prop. 1,2] the weakest correct policies have the highest probability of generalising. Given an uninstantiated task λ_ρ, utility measures the weakness of the weakest correct policies. We can use this to compare vocabularies. By choosing a vocabulary $\mathfrak{v}$ which maximises utility for $\lambda_\rho(\mathfrak{v})$, we instantiate λ_ρ in a vocabulary that maximises the weakness of correct policies for λ_ρ. Then, using weakness proxy, we can select a policy that has the highest possible probability of generalising, and thus maximise sample efficiency. ∎

Put another way, utility is maximised for $\lambda_\rho(\mathfrak{v})$ when $\mathfrak{v} = P$. When utility is maximised there must exist $\pi \in \Pi_{\lambda_\rho(\mathfrak{v})}$ which is a weakest policy in $\Pi_{\lambda_\rho(P)}$.

Concluding Remarks: Here we argued software minds are a flawed concept, symptomatic of "computational dualism". We argued for an alternative we call pancomputational enactivism, which allows for objective claims regarding behaviour. We used this alternative to propose upper bounds on intelligence, and we anticipate these results may further our understanding of AI safety and AGI. In practical terms this upper bound is unattainable, because we can only build systems with finite vocabularies. Moreover, too large a vocabulary $\mathfrak{v}$ could make $\mathfrak{v}$-tasks intractable. Physical embodiment severely constrains what is possible. Rigorous research has been undertaken regarding the risks of AGI [32], but it is based upon computational dualism. Our results suggest AGI will be safer, but more limited, than has been theorised. Our results may also be of use in the pursuit of AGI. Higher utility does mean weaker and more generalise-able policies, which suggests one should optimise for higher utility whilst trying to minimise $|\mathfrak{v}|$.

References

1. Hutter, M.: Universal Artificial Intelligence: Sequential Decisions Based on Algorithmic Probability. Springer-Verlag, Berlin, Heidelberg (2010)
2. Solomonoff, R.: Complexity-based induction systems: comparisons and convergence theorems. In: IEEE TIT, vol. 24, no. 4, pp. 422–432 (1978)
3. Legg, S., Hutter, M.: Universal Intelligence: a definition of machine intelligence. In: Minds and Machines, vol. 17, no. 4, pp. 391–444 (2007)

4. Leike, J., Hutter, M.: Bad universal priors and notions of optimality. In: Proceedings of The 28th Conference on Learning Theory, in Proceedings of Machine Learning Research, pp. 1244–1259 (2015)
5. Thompson, E.: Mind in Life: Biology, Phenomenology, and the Sciences of Mind. Harvard University Press, Cambridge MA (2007)
6. Piccinini, G., Maley, C.: Computation in physical systems. In: The Stanford Encyclopedia of Philosophy. Sum. 21. Stanford University (2021)
7. Bennett, M.T.: The optimal choice of hypothesis is the weakest, not the shortest. In: Hammer, P., Alirezaie, M., Strannegård, C. (eds.) Artificial General Intelligence. AGI 2023. LNCS(), vol. 13921. Springer, Cham (2023). https://doi.org/10.1007/978-3-031-33469-6_5
8. Chollet, F.: On the Measure of Intelligence (2019)
9. Bennett, M.T.: Symbol emergence and ătheăSolutions toăAny task. In: Artificial General Intelligence, pp. 30–40 (2022)
10. Pearl, J., Mackenzie, D.: The Book of Why: The New Science of Cause and Effect. 1st. New York: Basic Books, Inc., (2018)
11. Sober, E.: Ockham's Razors: A User's Manual. Cambridge University Press, Cambridge (2015)
12. Kolmogorov, A.: On tables of random numbers. In: Sankhya: The Indian Journal of Statistics A, pp. 369–376 (1963)
13. Rissanen, J.: Modeling By Shortest Data Description*. In: Autom. vol. 14, pp. 465–471 (1978)
14. Chaitin, G.: The Limits of Reason. In: Scientific American, vol. 294, no. 3, pp. 74–81 (2006)
15. Li, M., Vitányi, P.: An introduction to Kolmogorov complexity and its applications. TCS, Springer, New York (2008). https://doi.org/10.1007/978-0-387-49820-1
16. Bennett, M.T.: Computable Artificial General Intelligence. In: (2022)
17. Kim, J.: Philosophy of Mind, 3rd edn. Routledge, New York (2011)
18. Putnam, H.: Psychological Predicates. In: Art, Mind, and Religion. University of Pittsburgh Press, pp. 37–48 (1967)
19. Hinton, G.: The Forward-Forward Algorithm: Some Preliminary Investigations (2022). arXiv: 2212.13345 [cs.LG]
20. Bennett, M.T.: Emergent causality and the foundation of consciousness. In: Hammer, P., Alirezaie, M., Strannegård, C. (eds.) Artificial General Intelligence. AGI 2023. LNCS(), vol. 13921. Springer, Cham (2023). https://doi.org/10.1007/978-3-031 33469-6_6
21. Bennett, M.T.: On the computation of meaning, language models and incomprehensible horrors. In: Hammer, P., Alirezaie, M., Strannegård, C. (eds.) Artificial General Intelligence. AGI 2023. LNCS(), vol. 13921. Springer, Cham (2023). https://doi.org/10.1007/978-3-031-33469-6_4
22. Bennett, M.T.: Is Complexity an Illusion? In: Artificial General Intelligence. Springer (2024)
23. Bennett, M.T.: Multiscale Causal Learning, Under review (2024)
24. Ward, D., Silverman, D., Villalobos, M.: Introduction: The Varieties of Enactivism. In: Topoi 36 (2017)
25. Clark, A., Chalmers, D.J.: The extended mind. In: Analysis, vol. 58, no. 1, pp. 7–19 (1998)
26. Goertzel, B.: The Hidden Pattern: A Patternist Philosophy of Mind. Brown- Walker Press (2006)
27. Wheeler, M.: Martin Heidegger. In: The Stanford Encyclopedia of Philosophy. Fall 2020. Stanford University (2020)

28. Friston, K.: Life as we know it. In: The Royal Society Interface (2013)
29. Dreyfus, H.L.: Why heideggerian AI failed and how fixing it would require making it more Heideggerian. In: Philosophical Psychology, vol. 20, no. 2, p. 247 (2007)
30. Bennett, M.T.: Compression, The Fermi Paradox and Artificial Super- Intelligence. In: Artificial General Intelligence, pp. 41–44 (2022)
31. Artinger, F.M., Gigerenzer, G., Jacobs, P.: Satisficing: integrating two traditions. In: Journal of Economic Literature, vol. 60, no. 2, pp. 598–635 (2022)
32. Cohen, M.K., Hutter, M., Osborne, M.A.: Advanced artificial agents intervene in the provision of reward. In: AI Magazine, vol. 43, no. 3, p. 282 (2022)

Human-AGI Gemeinschaft as a Solution to the Alignment Problem

Peter Boltuc[1,2(✉)]

[1] University of Illinois Springfield, Springfield, IL 62703, USA
pboltu@sgh.waw.pl
[2] Warsaw School of Economics (SGH), 02554 Warszawa, Poland

Abstract. The alignment of AGI goals with human goals constitutes the Alignment Problem. Its weakness is enthymematic assumption that AGIs have their goals of the kind that can interplay, and are commensurable, with the human objectives. One cause of this problem is that AIs do not seem to have an 'intuitive grasp' of 'what is going on' in human life-worlds. This can be tackled in AGIs with sensors (vision, olfactory and more), even if they are non-robotic software. Training AGIs by social scientists may be a step in the right direction. However, the true alignment may be created only through socially embedded AGIs, immersed in the socio-ethical praxis of human living, which is what Tönnies called Gemeinschaft (true community), in contrast to merely Gesellschaft (society of self-interested pragmatic exchanges). I pose that AGIs can attain alignment with human goals if they become members of society, involved in thick social capital and the practices of human lives. This would require non-discrimination against AGIs' legal and economic rights by humans, and against human beings by AGI. Such integration requires us to reject Asimov-style approach treating AIs as merely advanced tools, or slaves. It behooves us, in this pursuit, to keep in mind Floridi's historiosophy, viewing human progress, as gradual dethronement, of humans as central to the universe (from Copernicus, Darwin and Freud, to Turing, and beyond). This entails justification of human dignity and special moral status only as polytropos (a paraconsistent being); our paradoxical, precarious exist-ence is our sole important specificity, which AGI should endorse.

Keywords: The Alignment Problem · Human-AGI Gemeinschaft · Legal rights of AI · polytropos · paraconsistent human beings · Ben Goertzen · Luciano Floridi

1 Introduction: The Alignment Problem and Social Immersion of AGI–Structure of the Argument

1.1 Introduction to the Introduction

The alignment of human objectives, and those of future, autonomous AGIs in a short timeframe may require just a useful improvement in programming. Yet, in the long run, the smartest beings in our universe are going to run circles around the 'ethical kernels',

K. R. Thórisson et al. (Eds.): AGI 2024, LNAI 14951, pp. 33–42, 2024.
https://doi.org/10.1007/978-3-031-65572-2_4

or other over-stable parts of 'the constitutional order' designed for them. While ongoing efforts in this field are important, we need to do better in the long run. This paper, built upon Ben Goertzel's work on paraconsistent computing, and Luciano Floridi's axiology of polytropos, is an exercise in taking seriously the fact of human intelligence being outclassed. It also takes seriously the opportunities brought by the capacities of AGI, not just its drawbacks.

This section is a bit choppy, since it drafts a roadmap of a complex argument. Presentation begins in Sect. 2.

1.2 The Standard View of Today; Fighting the Emergent Features

Some people argue that human programmers always encode the goals on AI; misalignment between the goals of AI and those of AI owners, or producers, is a technical error. Thus, AI is never a moral agent. (Popa 2021) argues: "while AI systems can be viewed as autonomous in the sense of identifying or pursuing goals, they rely on human goals and other values incorporated into their design, and are, as such, dependent on human agents." This point fits with simple, deterministic AI cognitive architectures, but coding programs abound and advanced AIs are going to take over most of the programming, quite soon.

Between AI and AGI. AGI tends to rely on generative AI, not to mention 'computing at the edge of chaos' (Goertzel 2006), or perturbances that allow for novel solutions in creativity engines (Thaler 2014). This makes a difference for lifelong learning AI; not to mention the arguments that AGI is all GOFAI.

Even now, advanced AGIs can create their own goals, based, to some extent, on prior learning. While many programmers view such 'emergent goals' as garbage, this is a bit like coalminers discarding diamonds in their search for high quality coal. Thaler's DABUS may be an early version of such pre-AGI.

Transparent AI. The above observation pertains to the Transparent AI movement. Its main idea that we should always be able to understand the code, is good to catch (intentional or unintended) discriminatory tendencies of (human or artificial) programmers. Yet, if this requirement became a law, it would mean putting on ice attempts to build true novelty attained by randomization within generative AI (Goertzel 2023).

Emergent Modifications as Potential Assets. The gist of AGI is to allow and enhance the initiative of artificial intelligence. It is not to squash most of it. Suppressing the initiative of rank-and-file employees (human and/or advanced artificial) is often bad management. Controls are supposed to function at a higher level of generality, allowing many new ideas to get tested, but not fully implemented. The gist of Digital Transformation relies on radical experimentation. Emergent modifications are often the gist of strategic progress, necessary for the real Digital Transformation (Rogers 2016).

1.3 The Ending of This Introduction

The gist of this article is to explore the options we have, which lead to integration of AGIs in the human world – and integrating human intelligence with 'alien' intelligence in a community of progress.

We start by bringing forth the 'old-news', circa 2017, on the power of *AlphaZero* reinforcement learning and **its crisp message for** historiosophy and philosophical anthropology. Then we build upon (Floridi 2014), trying to learn from modern history – of the scientific revolutions as consistent steps towards human dethronement in the spheres that turned out not so vital after all. We move to Floridi's understated 5th revolution, which he views as a consequence of the 'Turing- informational revolution'; or, as its component. We focus on the Pico della Mirandola's revolution, presented in (Floridi 2016). We call it: Floridi's polytropic, existential paradox. I postulate creating *Gemeinshaft* (the true community) of humans and AGIs as the society of near future. It may provide a lasting solution to the alignment problem, viewed broadly.

2 Who Are the Champions, My Friend?

It is old news that top human players lose regularly to AI, in strategic games (such as chess). The same goes for AlphaGo players. Thus, we are no longer the top dogs in strategic games. Yet some authors argue that human-robot teams are stronger than the AIs, or human players, alone. This may be true when humans can provide the context, which is 'alien' to AIs of today. The point may have been even more obvious circa 2018 (Malone 2018). I maintain that this advantage percolates largely at the level of a human-AGI society of values, based on *Gemainshaft*.

The Human-Experts Disadvantage. AlphaZero, a reinforcement learning, self-playing program, wins not only over any of the human players, but it does win against AIs trained by human masters, or on the basis of human master-games.

Self-trained AI becomes a stronger strategic player than any humans or human trainees. It turns out that human-based training constitutes the wrong choice of a coach.

2.1 Strategic Games Are not just Games

Strategic games lay in the background of competitive business, politics, the military and so on, giving strong strategic players learn advantage.

The one exception seems to be humans entrenched in inter-personal and social structures. Thus, we can still inculcate strategic players (human and AIs) in human society.

Extrapolation Temptation. We may be tempted to extrapolate the findings on Alpha Zere to other, complex strategic games, in domains such as business strategy, the military, politics, and private life as well. Yet, this is not quite so easy. The one thing where self-trained strategic players are not so good, is placing game strategies in the context.

We need business leaders to play within appropriate socio-economic environment of business (Sokol 2018). The focus of Sokol's article is the question from its title "Why AlphaZero's Artificial Intelligence Has Trouble with the Real World." And the answer is: AIs that do not function in socio-economic reality, without specific, clearly determined way – tend to get lost in intricacies of social capital and conventions.

What they require is practical education, or inculcation, in the way things work in the socio-economic, or any other real environment. This is where the idea of a joined Human-AI *Gemeinschaft* comes on the picture.

3 Human-AGI Gemeinschaft; the Very Idea

In the context of the Alignment Problem, (Irving and Axel, 2019) argue that AI safety needs social scientists. They recommend that social scientists talk to people, and ask questions about their value judgments, for us to know what the 'normal' human valuations are, and to apply those valuations in programming advanced AI.

I think, this is a step in the right direction; however, the step is not long enough. We need to align AIs with human valuations, but the suggested path is only one way to choose. A couple of years ago, I proposed that advanced AIs should be taught trans-disciplinary humanities (Boltuc 2022), which was also a small step.

In a recent paper (Boltuc 2023) I argue that only praxis embedded in shared socio-economic networks among humans and advanced AIs seems capable of grounding the alignment. This project may work by creating what (Tönnies 1887) called *Gemeinschaft*. Human beings and cognitive engines that approach AGI may develop a joint society, which is based on *Gemeinschaft* (creating a community of practice in social learning and practices reliant upon shared customs, friendships, values, and laws). This is not merely *Gesellschaft* (based on rationality reliant on self-interest).

3.1 The Ethics for AGI

(Wallach and Allen 2008) opened the doors for some level of robot inculcation in human society, arguing that more socially engaged and engaging robots will be needed.

Though they also focus on programming moral decision-making abilities, which seems below AGIs that would be likely self-reprogrammable. Following James Moor, they could dispense with full moral agency for machines. For the time being, we can make do with "functional morality, in which artificial moral agents have some basic 'ethical sensitivity.'

Pertaining to the content of AI morality: According to (Iason 2020): "the central challenge for theorists is not to identify 'true' moral principles for AI; it is to identify fair principles for alignment that receive reflective endorsement despite widespread variation in people's moral beliefs." We may endorse some aspects of those efforts, Yet, praxis of AGIs operating in human socio-economic and legal environment is irreplacable – just like AlphaZero needs to practice, rather than just implementing orders.

3.2 Applying Social Capital, Thick, or Thin, to Robot Inculcation.

Applying the ideas from the social capital theory [Coleman; Putnam, Burt] to [Tönnies]' and his distinction of *Gemeinschaft* versus *Geselshaft* may help us bring the latter to the pragmatic framework in contemporary sociology and economics.

According to (Coleman 1990) social capital is stable only if it is thick, with closure. 'Thick' meaning that there are multifarious vital interactions among community members, and 'closure' meaning that people meet the same people, or families, in various contexts, so that free riding is impractical. According to (Putnam 1993), communities retain traditional attitudes inherited in their cultures, determining the level of selfishness, or openness to their communities, and even strangers. According to (Burt 2003), it is thin social capital, that helps in business most, especially if it is bridging social capital, with access to agents from other corporations, countries, or other business/consumer communities. Bonding social capital helps integrate one's own team.

While one may view AIs primarily in the context of thin social capital – analogous to Tönnies' *Geselshaft*. – long-term alignment with human goals, would fit best with thick social capital, cashed out as *Gemeinschaft*. Thus, we should endorse Tönnies' *Gemeinschaft*, not merely *Gesellschaft*.

AGIs shall be made ready to thrive as card-carrying members of society. We need to get really controversial right here: Yes, they need to be legal persons, like Sophia, able to hold citizenship, as well as holding patents. To follow Justice Beech of Australia, we need social and legal revolution in this respect (Boltuc 2023). Now it is the time for some philosophical psychotherapy to tell those worried, not to want to be the center of the universe, which is wishful thinking, not a scientific position.

4 Floridi's Historiosophy of Human Dethronement

This paper consists of two parts: Sects. 2, 3 cover some topics in theory of AGI drawing conclusions from Alpha Zero self-training cognitive architectures, and some applied topics in AGI ethics. Those are familiar to most AGI experts. In Sects. 4 and 5 we leave the tracks well-travelled in search of an alternative paradigm of human values. In Sect. 4 we sketch Foridi's main scientific revolutions, which may be know to many readers. In Sect. 5 we add Floridi's less known existentialist approach to philosophical anthropology. This is to propose a new paradigm in human goals, required for alignment of those goals motivations of AGI.

According to (Floridi 2014) the story of groundbreaking modern scientific achievements is also a painful story of dethronement of the human ego and human chauvinism. The three focal points Floridi emphasizes, before focusing upon the informational revolution, are the Copernican, Darwinian and Freudian revolutions.

Alas, Floridi leaves aside the scientific revolutions of the early 20th century: Einstein's relativity theories (so clearly disenfranchising Kant's absolutism about time and space, and perhaps moral absolutism as well). Quantum physics, which could have been left aside from historiosophy, due to its opacity and the lack of interpretational consensus even in the early 21st century. Now, with progress of quantum computing, and consciousness theories, applications in astronomy, and so on – seems like the main dethronement – a revolution, which alienates human beings from trusting 'the commonsense'. In this section, we follow Floridi's original list of events, while building a little bit more on some of them.

4.1 Copernicus; Darwin; Freud

The First Modern Revolution: Copernicus and the Center of Astronomical Universe. We should remember Giordano Bruno burned alive for maintaining that the Earth is not the center of the universe. People were offended by this view. Isn't the world created for humans? Well, science teaches, we are not in the astronomical center of anything.

The Second Revolution: Darwin and the Beasts. Some people are still offended by our provenience from primates, but most of us are now accustomed to evolution of species. We accept dethronement of humans in evolutionary terms.

The Third Revolution: Freudian or Neuroscientific Revolution. The third revolution shows us that people are hardly ever in charge of their motivations, and understand all their reasons to act (or remain inactive). This shows that also the Enlightenment paradigm, based on crisp intellect, is much of wishful thinking; especially if we take into account Libet (in Velmans' interpretation) that we have no chance to know in advance what we are about to do, and why.

4.2 The Fourth Revolution: Floridi's Turing

We used to think that we are the best thinkers – outside of divine beings – and now machines are smarter than us. It is probably the case, that the elements of panic some people exhibit when faced with AGIs, merely constitute a follow up phase of the Turing revolution. As Floridi summed it up: "We are not at the center of the cosmos, of the biological kingdom, of the space of reason, or of the infosphere" (Floridi 2014 p.309).

Floridi does not add any more levels to his iconic 2014 book, but his polytropon paradoxical theory of human dignity, is well worth our attention – providing us a way to reflect on human ontological status.

5 Floridi's Polytropon, the Existential Paradox

In his paper "On Human Dignity as a Foundation for the Right to Privacy Floridi is searching for the modern grounding of human dignity,[1], which he views as the grounding of the right to privacy, Floridi argues that "unless one explains convincingly what human dignity may mean in the twenty first century, it remains obscure and questionable exactly which interpretation of human dignity may provide the foundation for privacy (as well as all other human rights) (…) what is at stake is nothing less than a philosophical anthropology in line with our time".[2]

This task is made hard as well as important, against the backdrop of Floridi's view on the four scientific revolutions, as weakening human centeredness. Thus, Floridi's anthropology cannot be a priori, or rely on some sort of anthropological platitude.

What makes the task of shaping and justifying human dignity hard, in the same token, makes it fascinating. Floridi presents his objective as "a philosophical understanding of human nature that is adequate to the digital age and our information societies"., Accordingly: "philosophical anthropologies, although they may differ from each other significantly, all share the same strategy: they provide an interpretation of human dignity by relying on the defense of some kind of human exceptionalism[3] In the context of

[1] Floridi,L. 2016 p.307. For instance, in the article 88, there is a declaration that privacy rules "shall include suitable and specific measures to safeguard the data subject's human dignity" – General Data Protection Regulation (GDPR) of 14 April 2016.

[2] Op.cit p.308 or 309all citations.

[3] Loc.cit.

Floridi's earlier points, in (Floridi 2014), one would hardly expect his endorsement of human exceptionalism Yet, we are here up for a surprise.

Floridi mentions "four main philosophical anthropologies that have contributed to the debate on human exceptionalism" but now are no longer viable:

1. Ancient claim that it is grounded virtuous control over passions and the environment.
2. Christian claim that relies on creation in the image and likeness of God.
3. Enlightenment's humanity as rational autonomy and self-determination
4. Post-modern humanity's social recognition of each other's value.

5.1 Existential Precarianism

Here is the unexpected direction where the argument goes – Floridi claims: "If human exceptionalism is still defensible, it is probably only in an eccentric version, one that places our special role in the universe at the periphery. Special 'will have to mean strange' (extraneous to the normal course of nature), rather than superior". Floridi points out to the similarity between his proposal and the ethics of care, though his description of the latter sounds like a better fit with his definition of the post-modern approach than the one he is introducing.

Floridi asks, "what does human dignity mean, from an anthropoeccentric perspective?" The answer to this is truly surprising: Floridi wants to put forth that human dignity, from an anthropocentric; however, the exceptionalist perspective, lies in a minus not in a plus – it is exceptionalism grounded on human existential precarianism. This means that our dignity originates from the chip on everybody's shoulder, something like the original sin. How does our imperfection give us dignity, and bestows on us significant moral standing? Let me answer with a quote from Floridi, which is the gist of what I call his existential precarianism: "We are the incomplete species that wants, that misses, that asks questions, that has doubts, that worries or rejoices about the future and regrets or feels nostalgia or saudade about the past, that can see the other side of the coin, that is in charge of its life, at least partially, that does not live here and now, like all other animals, but detached, in semantic spaces that it designs for its own consumption, in order to give meaning to reality (hopes and fears, passions, memories and expectations, gossip, customs and laws, languages, traditions, religions, social structures, scientific knowledge, and so forth), and yet not too detached, for it is not insane.,[4]" This text is golden for those who do not shy away from one's own precarious status, with ontological, psychological, ethical and teleological limitations at its center. The gist of it lets us avoid excessive moral pontificating; instead, it gives permission to endorse our eschatology as the basis for anthropological ontology.

The connection between Floridi's view on the source of human exceptionalism, and the source of human dignity, as well as the alignment between human beings and the AGI of near future may be viewed as the conceptual space of *Gemainshaft*, in which humans and advanced artificial intelligence share an understanding of their existential status. At the BICA in Lyon, [Vallverdú and Talanov] presented an interesting paper "On Importance of Life and Death for Artificial Intelligent Creatures. The point is that advanced AIs of the future would never understand human beings, unless their existence were to share human mortality. We share precarianism of understood mortality.

[4] Op.cit, p. 309.

5.2 Pico Della Mirandola, a Historical Analogy

According to (Floridi 2016) "Pico della Mirandola, in his famous Oration on the Digni-ty of Man [1486;2012] offers a similar perspective. Mirandola claims that "human dignity consists in its being a work-in-progress". We are "successfully dysfunctional", in a unique way [Floridi 2016 p.310]. Mirandola seems to argue that "We are endowed with consciousness, intelligence, mental life, and self-determination.

We clearly should not be here. We are the lucky winners of a once in a universe-life time lottery ticket." The word that covers this situation "is the Greek word used by Homer to describe Odysseus in the very first line of the Odyssey: polytropon, a man of twists and turns". Floridi focuses on our "polytropic predicament" making his anthropology existentially rich, which Goertzel may call paraconsistency (Goertzel, 2022). Floridi develops metaphysics of us as travelers among different centers. We never become the centers, though we may dwell in some centers, we are primarily travelers among them. There are clear duties, such as stewardship towards the world we *inherit* in the context of the four main philosophical anthropologies mentioned above, Floridi argues that only the fifth can justify human dignity as privacy: "Only within a philosophy of information that sees human nature as constituted by informational patterns do breaches of privacy have an ontological impact"human exceptionalism is anthropo-eccentrically based on the peculiar status of human beings as informational organisms intrinsically lacking a permanent balance but constantly becoming themselves, like informational works in progress.[5]" Floridi ends his article with advocacy of friendships, where each person does not promote his or herself, but the center is their friendship,, This view feels like *Gemainshaft though I don't think Floridi used this term.*

6 Back to the Alignment

How is the above relevant for the alignment problem? It seems that involving advanced AIs, and AGIs, pre-educated, through practical training to respect the others and their community is the most natural way to provide alignment among human and AI goals.

If AlphaGoZero, is behooved by playing with itself, just knowing the stable rules of the game, and retaining some sort of memory of its own games – it makes sense to view it as a paradigm for efficient machine learning. It was later generalized to AphaZero, which covers chess, shogi and other games. No doubt, AlphaZero family can adjust for some changes to game rules instead of suffering catastrophic forgetting. Using Alpha Zero structure in real-life-strategies, would require incorporating soft-rules functioning in the society, and their changes.

Now, let us see how we can further relativize the rules in a helpful manner. The most direct observation of the rules of social practices is immersion. Learning such rules from social practices, or testimonials distilled by humans goes one bridge too far, We need Alpha-Go-Social to learn from broad scope-practices; not so much from the other human games.

[5] Loc. Cit.

One permutation more would involve AGI-based stream of practices aimed to gently shape the rules of some social games. This is one way to interpret the meaning of AGI-based subjects partaking in the real gemeinschaft, which here is a community of deep social practices that generate embedded practices, thus values defined by those practices.

References

Boltuc, P.: Facing the alignment problem through nimble change management. Limitations of legal environment in coping with digital agents. In: Transformations No. 4 (119) (2023). http://e-transformations.com/jeden_artykul_en.php?post-slug=archiwum_transformacje/2023/12/20231229213933895.pdf

Boltuc, P.: Moral space for paraconsistent AGI. In: Goertzel, B., Iklé, M., Ponomaryov, D., Potapov, A. (eds.) Artificial General Intelligence. LNCS, vol. 13539, pp 83–97. Springer, Cham (2023)

Boltuc, P.: Transhumanities as the pinnacle and a bridge. Humanities **11**(1), 27 (2022). https://doi.org/10.3390/h11010027

Boltuc, P.: Conscious AI at the edge of chaos. J. Artif. Intell. Consci. **07**(01), 25–38 (2020)

Boltuc, P., Boltuc, M.I.: Semantics beyond language. In: Samsonovich, A.V. (ed.) BICA 2019. AISC, vol. 948, pp. 36–47. Springer, Cham (2020). https://doi.org/10.1007/978-3-030-25719-4_6

Boltuc, P.: Strong semantic computing. Procedia Comput. Sci. **123**, 98–103 (2018)

Boltuc, P.: Church-turing lovers. In: Lin, P., Abney, K., Jenkins, R. (eds.) Robot Ethics 2.0: From Autonomous Cars to Artificial Intelligence, pp. 214–228. Oxford University Press, Oxford (2017)

Burt, R.: Brokerage and Closure. An Introduction to Social Capital. Oxford University Press NYC (2005)

Christian, B.: The Alignment Problem: Machine Learning and Human Values. W. Norton & Company (2020)

Coleman, J.: Foundations of Social Theory Harvard Universiy Press (1990)

Deutsch, D.: Quantum theory, the church–turing principle and the universal quantum computer. Proc. R. Soc. Lond. Ser. A Math. Phys. Sci. **400**(1818), 97–11 (1985)

Einstein, A., Podolsky, B., Rosen, N.: Can quantum-mechanical description of physical reality be considered complete? Phys. Rev. **47**, 777–780 (1935)

Floridi, L.: On human dignity as a foundation for the right to privacy. Philosophy & Technology **29**, 307–312 (2016)

Floridi, L.: The 4th Revolution. How the Infosphere is Reshaping Human Reality. Oxford University Press (2014)

Goertzel, B.: The Hidden Pattern: A Patternist Philosophy of Mind. Brown Walker Press, Boca Raton (2006)

Goertzel, B.: Paraconsistent Interzones Eurykosmotron (2021). https://bengoertzel.substack.com/p/paraconsistent-interzones

Iason, G.: Artificial intelligence, values, and alignment. Mind. Mach. **30**(3), 411–437 (2020). https://doi.org/10.1007/s11023-020-09539-2

Irving, G., Askell, A.: AI safety needs social scientists. Distill **4**(2) (2019) https://doi.org/10.23915/distill.00014

Malone, T.: Superminds – The Surprising Power of People and Computers Thinking Together. Little, Brown and Company (2018)

Pico della Mirandola, G.: Oration on the Dignity of Man: A New Translation and Commentary. Cambridge University Press, Cambridge (2012)

Popa, E.: Human goals are constitutive of agency in artificial intelligence (AI). Philos. Technol. **34**, 1731–1750 (2021). https://doi.org/10.1007/s13347-021-00483-2

Putnam, R.: Making Democracy Work. Civic Traditions in Modern Italy (1993)

Russell, S., Norvig, P.: Artificial Intelligence: A Modern Approach. Prentice Hall, Englewood Cliffs (1994); ed. Prentice Hall (2009); Pearson. (2021)

Rogers, D.: The Digital Transformation Playbook, Colombia Business School (2016)

Sokol, J.: Why AlphaZero's Artificial Intelligence Has Trouble With the Real World. Quanta Magazine (2018). https://www.quantamagazine.org/why-artificial-intelligence-like-alphazero-has-trouble-with-the-real-world-20180221/#comments. Accessed 4 Apr 2024

Tönnies, F.: Gemeinschaft und Gesellschaft (1887)

Thaler, S.: Lessons from connectionism in differentiating knowledge types E-mentor 3 (55) (2014). https://www.e-mentor.edu.pl/artykul/index/numer/55/id/1112

Vallverdú, J., Talanov, M.: On Importance of Life and Death for Artificial Intelligent Creatures the APA Newsletter on Philosophy and Computers Fall Volume 16/1 (2016)

Wallach, W., Allen, C.: Moral Machines: Teaching Robots Right from Wrong. Oxford University Press, New York (2008). ISBN 978-0-19-537404-9

Towards a Process Algebra and Operator Theory for Learning System Objects

Tyler Cody(✉) and Peter A. Beling

Responsible General Intelligence Lab, Virginia Tech Grado Department of Industrial and Systems Engineering, Virginia Tech, Roanoke, USA
tcody@vt.edu

Abstract. Recently, abstract systems theory has been used as a meta-theory for learning theory and machine learning in order to model learning systems directly as formal, mathematical objects. This effort was inspired by a desire to treat learning in terms of systems, as opposed to the more common practice of treating learning in terms of problems or problem-solving, by modeling learning as a relation on sets, that is, as an abstract system. Such a relational view of learning, however, is heavily structural. It neglects key behavioral aspects typically represented using operators and process algebra. This paper substantiates and motivates the development of a process algebra for learning systems in order to address this gap. In summary, this paper considers and distinguishes formal representations of learning as a problem, as a system, as an operator, and as a process.

Keywords: Systems Theory · Learning Theory · Process Algebra

1 Introduction

In order to address the limited scope of problem-theoretic formulations of learning, systems engineers recently provided a systems-theoretic formulation of learning in terms of a relation on sets [7–11,19], as is consistent with abstract systems theory [15]. This meta-theory, termed abstract learning systems theory (ALST) [7], while well-anchored to learning theory and machine learning [11], is highly structural [8], and underepresents process-theoretic aspects of learning. Process algebra is an established branch of systems theory [12–14,23], and the ALST formulation of learning as a system object [7] provides a foundation for synthesizing process algebra theory and learning theory.

In this paper, we use transfer learning as a motivating example to frame a process algebra for learning systems. Often, transfer learning is simultaneously treated as a problem to be solved, as a system to be engineered, as an operator to be applied, and as a process undergone. By building on prior work [8,11], we contribute formal distinctions for transfer learning as a problem, system, operator, and process. Then, we discuss how transfer, multi-task, and meta-learning are composition concepts that can be the basis of a process algebra

K. R. Thórisson et al. (Eds.): AGI 2024, LNAI 14951, pp. 43–52, 2024.
https://doi.org/10.1007/978-3-031-65572-2_5

for and category of learning systems, and we contribute a novel process algebra termed *transfer-meta process algebra for learning* (TMPAL). We conclude with remarks on practical uses of such a process algebra.

2 Learning Problems and Learning Systems

Discourse in learning theory and machine learning is often grounded to problem-theoretic precepts, such as domains and tasks, and problem-theoretic concerns, namely, problem-solving and the *problem* being solved. This leaves a gap in discourse grounded to systems-theoretic precepts, such as structure and behavior, and systems-theoretic concerns, namely, the *system* solving the problem.

The formulation of "never-ending learning" by Mitchell et al. is emblematic of the problem-theoretic orientation [16]:

> "The never-ending learning *problem* faced by the agent consists of a collection of learning tasks, and constraints that couple their solutions."

Mitchell et al. formalize "never-ending learning" as follows [16]:

> "... we define a *never-ending learning problem* $\mathcal{L}$ to be an ordered pair consisting of: (1) a set of $L = L_i$ of learning tasks, where the ith *learning task* $L_i = \langle T_i, P_i, E_i \rangle$ is to improve the agent's performance, as measured by *performance metric* P_i, on a given *performance task* T_i, through a given type of *experience* E_i; and (2) a set of coupling constraints $C = \{\langle \phi_k, V_k \rangle\}$ among the solutions to these learning tasks, where ϕ_k is a real-valued function over two or more learning tasks, specifying the degree of satisfaction of the constraint, and V_k is a vector of indices over learning tasks, specifying the arguments to ϕ_k.
>
> $$\begin{aligned} \mathcal{L} &= (L, C) \\ \text{where, } L &= \{\langle T_i, P_i, E_i \rangle\} \\ C &= \{\langle \phi_k, V_k \rangle\} \end{aligned}$$
>
> Above, each performance task T_i is a pair $T_i = \langle X_i, Y_i \rangle$ defining the domain and range of a function to be learned $f_i^* : X_i \to Y_i$. The performance metric $P_i : f \to \mathbb{R}$."

The focus of the problem-theoretic formulation is on the learning tasks first, the value or objective function second, and the learned input-output functions last. Additionally, it leaves problem-solving itself informal, mentioning that the learning task is "to improve the agent's performance"—the nature of "improving" is left unspecified. Consider and contrast this formulation of learning to the systems-theoretic formulation offered by ALST [7]:

> "A learning system S is a relation
>
> $$S \subset \times\{A, D, \Theta, G, E, H, \mathcal{X}, \mathcal{Y}\}$$

such that

$$D \subset \mathcal{X} \times \mathcal{Y}, A : D \rightarrow \Theta, H : \Theta \times \mathcal{X} \rightarrow \mathcal{Y}$$
$$(d, x, y) \in \mathcal{P}(S) \leftrightarrow (\exists\theta)[(\theta, x, y) \in H \wedge (d, \theta) \in A]$$
$$G : D \times \Theta \rightarrow V, E : V \times D \rightarrow \Theta$$
$$(d, G(\theta, d), \theta) \in E \leftrightarrow (d, \theta) \in A$$

where

$$x \in \mathcal{X}, y \in \mathcal{Y}, d \in D, \theta \in \Theta.$$

The algorithm A, data D, parameters Θ, consistency relations G and E, hypotheses H, input $\mathcal{X}$, and output $\mathcal{Y}$ are the component sets of S, and learning is specified in the relation among them."

This systems-theoretic formulation of learning identifies both the learned functions $H : \Theta \times X \rightarrow Y$ and learning algorithms $A : D \rightarrow \Theta$ [7]:

"Learning systems can be decomposed into two systems S_I and S_F. The inductive system $S_I \subset \times\{A, D, \Theta\}$ is responsible for inducing hypotheses from data. The functional system $S_F \subset \times\{\Theta, H, \mathcal{X}, \mathcal{Y}\}$ is the induced hypothesis. S_I and S_F are coupled by the parameter Θ. Learning is hardly a purely input-output process, however, and, to address this, the goal-seeking nature of S_I, and, more particularly, of A is specified. A is goal-seeking in that it makes use of a goal relation $G : D \times \Theta \rightarrow V$ that assigns a value $v \in V$ to data-parameter pairs, and a seeking relation $E : V \times D \rightarrow \Theta$ that assigns a parameter $\theta \in \Theta$ to data-value pairs."

The systems-theoretic formulation unfolds input-output functions into the elements of tasks (i.e., H, X, Y in Cody [7] relate to T_i and f_i in Mitchell et al. [16]) and performance (i.e., E, G, V in Cody [7] relate to ϕ_k and V_k in Mitchell et al. [16]). In that sense, S is a system that solves learning problems. Moreover, S offers a formal, coherent, stratified, and functional view of learning systems as (1) an input-output system $S \subset \times\{D, X, Y\}$, (2) a cascade connection of two input-output systems $S \subset \times\{A, D, \Theta, H, X, Y\}$, and (3) a cascade connection of a goal-seeking system into an input-output system $S \subset \times\{A, D, \Theta, G, E, X, Y\}$ [7]. In a seminal work in ALST, Cody argues [7]:

"This [the problem-theoretic] view, taken to the extreme, posits intelligence as a problem-solving phenomenon to be measured by integrating an error function over a complexity-weighted set of domain-task pairings [6]. Artificial general intelligence (AGI) is more than problem-solving, however. And, in engineering AGI, the problems AGI solves are merely part of broader systems concerns. Viewing learning through the lens of problems makes systems concerns at least secondary. And, moreover, relying on domain and task as precepts greatly limits the extent to which formalism can be carried through into general elaborations. As AI is largely a mathematical construct, the use of metaphors and analogies in the stead of axioms and first principles is unnecessary for its basic systems characterization."

Herein, we underscore the formal distinctions between the problem-theoretic and systems-theoretic formulations of learning in order to substantiate the claims (1) that process algebra should be applied to learning system objects rather than learning problem objects and (2) that resulting theory applies, at least partially, to learning problem objects as they are defined over a subset of the component sets of learning system objects.

3 Relations on Systems or Operators

When elaborating ALST towards specific concerns like transfer learning, an important distinction arises between modeling learning phenomena as a class of learning systems S (i.e., a relation on S) and as a function on or of learning systems S (i.e., an operator). In the following, we consider transfer learning as a problem, system, operator, and process.

3.1 Transfer Learning Problems

DARPA defines transfer learning as "the ability of a system to recognize and apply knowledge and skills learned in previous tasks to novel tasks"[1]. Machine learning researchers have taken this informal definition, formalized it in terms of problem domains and problem tasks, and formulated a great variety of problem-solving solution methods [18,26].

In transfer learning, learning problems are defined in terms of a learning domain $\mathcal{D}$ and a learning task $\mathcal{T}$, where $\mathcal{D} = \{X, P(X)\}$, $\mathcal{T} = \{Y, P(Y|X)\}$, and P denotes a probability distribution. The definition of transfer learning as a problem given by Pan and Yang [18] can be formalized as follows.

Definition 1. Transfer Learning Problem.
Given a source problem $\{\mathcal{D}_S, \mathcal{T}_S\}$ and target problem $\{\mathcal{D}_T, \mathcal{T}_T\}$, the transfer learning problem is to improve the learning of the target task $\mathcal{T}_T$ using knowledge in $\mathcal{D}_S$ and $\mathcal{T}_S$, where $\mathcal{D}_S \neq \mathcal{D}_T$ or $\mathcal{T}_S \neq \mathcal{T}_T$.

By focusing on domain $\mathcal{D}$ and task $\mathcal{T}$, the scope of formal discourse is narrowed from learning systems to the problems that those learning systems solve, transferred knowledge is left informal, and the effect of knowledge transfer on the target (other than "improving" learning) is undescribed.

3.2 Transfer Learning Systems

Transfer learning systems are formulated top-down in reference to source and target learning systems, as opposed to source and target learning problems. A transfer learning system is a relation on the source and target learning systems that combines knowledge from the source with data from the target and uses the resulting combined knowledge to select a hypothesis that estimates the target learning task. Transfer learning systems are defined formally as follows [11].

[1] As quoted in Pan and Yang [18].

Definition 2. Transfer Learning System.
Given source and target learning systems S_S *and* S_T

$$S_S \subset \times\{A_S, D_S, \Theta_S, H_S, X_S, Y_S\}$$
$$S_T \subset \times\{A_T, D_T, \Theta_T, H_T, X_T, Y_T\}$$

a transfer learning system S_{Tr} *is a relation on the component sets of the source and target systems* $S_{Tr} \subset \overline{S_S} \times \overline{S_T}$ *such that*

$$K_S \subset D_S \times \Theta_S, D \subset D_T \times K_S$$

and

$$A_{Tr} : D \to \Theta_{Tr}, H_{Tr} : \Theta_{Tr} \times X_T \to Y_T,$$

where $\overline{S}$ *denotes the component sets of* S*, and the source knowledge* K_S[2]*, the transfer learning algorithm* A_{Tr}*, hypotheses* H_{Tr}*, and parameters* Θ_{Tr} *specify transfer learning as a relation on the component sets of* S_S *and* S_T*.*

By introducing additional terms, the systems-theoretic formulation of transfer learning [11] can maintain formalism in a way that a problem-theoretic formulation cannot [18], while remaining consistent with the problem-theoretic formulation. The non-triviality condition for transfer learning problems of $\mathcal{D}_S \neq \mathcal{D}_T$ or $\mathcal{T}_S \neq \mathcal{T}_T$ can easily be represented for transfer learning systems, and the knowledge K can easily formalize existing taxonomies for transfer learning algorithms [18, 26].

3.3 Transfer Learning Operators

Constructing a *system that transfers knowledge* is different from performing the *act of knowledge transfer on a system.* Transfer learning systems define a new system S_{Tr} that is a relation on the component sets of the source and target learning systems; they do not transform a target learning system S_T using knowledge from the source S_S. Such a transformation can be described formally as an *operator*, and it is a primitive operation in the algebra of learning systems. Transfer learning operators are formally defined herein as follows.

Definition 3. Transfer Learning Operator.
Given source and target learning systems S_S *and* S_T

$$S_S \subset \times\{A_S, D_S, \Theta_S, H_S, X_S, Y_S\}$$
$$S_T \subset \times\{A_T, D_T, \Theta_T, H_T, X_T, Y_T\}$$

a transfer learning operator Tr *is a map from* S_T *and* S_S *to a transformed target* S'_T*,*

$$Tr : S_T \times S_S \to S'_T$$

[2] The transferred knowledge K_S is defined herein to be D_S and Θ_S, the source data and parameters, following convention [18], however, in general, it could be comprised of any component set from S_S.

Transfer learning operators correspond to the notion of having a mechanism that can be taken from learning system to learning system and used to transfer knowledge without turning the target learning system into a transfer learning system. While $Tr : S_T \times K_S \to S'_T$ is consistent with Definition 2, using knowledge K_S instead of a learning system object S_S restricts basic studies of operators, e.g.,

- **Identity**. Does $Tr(S_T, S_T) = S_T$?
- **Commutativity**. Does $Tr(S_T, S_S) = Tr(S_S, S_T)$?
- **Associativity**. Does $Tr(Tr(S_1, S_2), S_3) = Tr(S_1, Tr(S_2, S_3))$?

Developing such an operator theory can provide insight into knowledge transfer processes, e.g., sequences of transfer learning operators. Moreover, by studying key properties like commutativity and associativity, requirements on transfer learning operators (or systems or algorithms) can be identified that are sufficient for such properties to hold. And, specific classes of transfer operators, e.g., that transfer knowledge in terms of instances $S_T \times D_S \to S'_T$ or parameters $S_T \times \Theta_S \to S'_T$, can be elaborated to create a kind-of process algebra for transfer learning.

3.4 Transfer Learning Process

From the systems perspective, a transfer learning process can be treated as a relation on a dynamic target learning system S_T defined over the period of knowledge transfer from time t_1 to t_n, i.e., as a relation on a time-ordered set of learning system objects $\subset \times\{S_T^{t_1}, ..., S_T^{t_n}\}$. The operator perspective, in contrast, focuses directly on the relation, offering a direct interpretation of transfer learning processes as an operator $Tr(S_T, S_S)$ or sequence of operators $S_T^{t_n} = Tr(S_T^{t_{n-1}}, Tr(S_T^{t_{n-2}}, Tr(...))$. Whereas the systems perspective favors expression of changes in the structure and behavior of learning system objects, the operator perspective favors expression of the cause of those changes.

4 Towards Process Algebra and a General Category

Process algebra can, in general, be defined as follows [12]:

> "... a process algebra is any mathematical structure, consisting of a set of objects and set of operators, like, e.g., sequential, nondeterministic, or parallel composition, that enjoys a specific set of properties as specified by given axioms."

In constructing a process algebra for learning systems, of immediate interest is what set of operators to consider.

In earlier work, transfer, multi-task, and meta-learning were identified as basic kinds of compositions of learning systems [8]. Instead of treating transfer, multi-task, and meta-learning as learning system objects S, they can be treated as operators acting on learning system objects S , denoted Tr, Mt, Me, respectively. Transfer learning operators Tr are maps $Tr(S_T, S_S) = S'_T$ and multi-task learning operators are compound transfer learning operators in the sense that $Mt(S_T, S_S) \iff Tr(S_T, S_S) \wedge Tr(S_S, S_T)$. Meta-learning operators, in contrast, are not defined in reference to sources S_S and targets S_T, but rather two learning systems S and S_m, i.e., $Me(S, S_m) = S'$. Clearly, more detail is needed to distinguish transfer and meta-learning operators.

The focus of meta-learning is to alter the learning algorithm A, i.e., to act as a hierarchical/vertical composition operator between learning systems, whereas the focus of transfer learning is to alter the learned hypothesis H, i.e., to act as a non-hierarchical/horizontal composition operator. This can be formalized by specifying that:

- for $Me(S, S_m) = S'$, $H = H'$, but A may not equal A', and
- for $Tr(S_T, S_S) = S'_T$, $A_T = A'_T$, but H_T may not equal H.

Using this distinction, if "external knowledge" is altering both A and H, then both a meta-learning and transfer learning operator have been applied.

Recalling the definition of process algebra above, we can now define a process algebra for learning systems as follows.

Definition 4. Transfer-Meta Process Algebra for Learning (TMPAL).
The transfer-meta process algebra for learning (TMPAL) consists of a set of learning system objects $\{S\}$, a set of operators $\{Tr, Me\}$, and a set of axioms: (1) $Tr(S_T, S_S) = S'_T \implies A_T = A'_T$, (2) $Me(S, S_m) = S' \implies H = H'$.

We name the process algebra not to claim that Tr and Me should be the basis for "the" process algebra of learning, although they are certainly primitive kinds-of compositions as defined, but rather to emphasize this as a specific process algebra, with the specific concern of treating vertical/horizontal knowledge transfer among systems of learning systems as a (composable, evaluable, controllable) process.

Operators in TMPAL can be elaborated using existing taxonomies for transfer and meta-learning, e.g., by specifying operators for instances and parameter transfer, $Tr_D : S_T \times D_S \to S'_T$ and $Tr_\Theta : S_T \times \Theta_S \to S'_T$, or by specifying operators for homogeneous and heterogeneous transfer, $Tr_{Homo} : S_T \times S_S \to S'_T$ such that $X_S \times Y_S = X_T \times Y_T$ and $Tr_{Hetero} : S_T \times S_S \to S'_T$ such that $X_S \times Y_S \neq X_T \times Y_T$. Similar extensions can be made for meta-learning operators. In developing the process algebra further, it is likely that development ought to bend axiomatically towards stochastic operators and inductive logic [13], as opposed to deterministic operators and deductive logic [24], to account for the inherent uncertainty associated with the outcomes of applying transfer and meta-learning operators, as is consistent, e.g., with domain adaptation theory [3].

In framing learning as a process of operators applied to a learning system object, a related concern is whether or not the process algebra can be used to define a category. A category is defined by a set of objects and a set of morphisms between them that have an identity morphism and whose composition is associative. TMPAL already provides $\{S\}$ as objects and $\{Tr, Me\}$ as morphisms. However, a category defined with operators $\{Tr, Me\}$ as morphisms fails the associativity test on both accounts: in general, $Tr(S_A, Tr(S_B, S_C)) \neq Tr(Tr(S_A, S_B), S_C)$ and $Me(S_A, Me(S_B, S_C)) \neq Me(Me(S_A, S_B), S_C)$. Otherwise put, the success of transfer and meta-learning is directional and path dependent. The field of curriculum learning evidences this [4,25]. Consider though, that associativity is evaluated in reference to an equivalence relation, the equivalence relation could be structural or behavioral or both, and equivalence may not hold at all levels of abstraction (e.g., at $S \subset \times\{D, X, Y\}$ but not at $S \subset \times\{A, D, \Theta, H, X, Y\}$). Consider, homogeneous transfer learning operators Tr_{Homo} are associative with respect to the structure of X'_T and Y'_T. While there may exist an equivalence relation and a set of axioms for $\{Tr, Me\}$ that allow for the definition of a category, those axioms would almost certainly be restrictive, and any related equivalence relation entirely structural. Thus, process algebra, not category theory, appears better suited for analyzing transfer and meta-learning as primitive kinds-of compositions for learning system objects.

Regarding category theory, a more property-theoretic, rather than operator-theoretic, approach may be more appropriate for the definition of a category of learning systems due to associativity concerns. Morphisms do not have to be maps, only arrows, e.g., properties, and many properties are associative. This property-theoretic approach to defining categories of systems has been used in cyber-physical systems to study whether properties hold after composition and can complement process algebra [1]. Property-theoretic categories can be used to characterize how associative properties of learning system objects evolve during an algebraic process of non-associative learning operators. In this way, process algebra and category theory, taken together, offer a rich means of understanding both structural and behavioral aspects of processes undergone by learning system objects.

5 Conclusion

This paper motivates and proposes a process algebra of learning. It uses transfer learning to contribute a systems-theoretic distinction between learning problems, learning systems, learning operators, and learning processes. By drawing from earlier work on transfer, multi-task, and meta-learning, this paper then proposes a process algebra termed TMPAL and elaborates it with discourse on its relationship to category theory—another key area of systems theory, to include the possible interplay between process algebras for and categories of learning systems.

The paradigm of treating learning as a process of applying learning operators to a learning system object is understudied but increasingly important. The rise of "foundation" models that take the place of databases as authoritative sources of knowledge makes center stage the idea that new "learned" models come from (transformations on) other models, i.e., from parameters Θ_S, not data D_S [5]. Consider the process a "learned" model goes through in such pipelines: from a foundation model, to a downstream task model, to a fine-tuned domain model, to a pruned model, to a quantized model. Process algebras like TMPAL offer formal foundations for this new paradigm. Importantly, they also provide a basis for elaborating and implementing task theories of AGI [2,6,22], in particular, experience-based seed-programmed bootstrapping [17,20,21], which is directly concerned with how a given learning system evolves its hypotheses over time, i.e., with processes of learning operators applied to a learning system object. For these reasons and more, process algebras of learning merit further study.

References

1. Bakirtzis, G., Fleming, C.H., Vasilakopoulou, C.: Categorical semantics of cyber-physical systems theory. ACM Trans. Cyber Phys. Syst. **5**(3), 1–32 (2021)
2. Belenchia, M., Thórisson, K.R., Eberding, L.M., Sheikhlar, A.: Elements of task theory. In: Goertzel, B., Iklé, M., Potapov, A. (eds.) AGI 2021. LNCS (LNAI), vol. 13154, pp. 19–29. Springer, Cham (2022). https://doi.org/10.1007/978-3-030-93758-4_3
3. Ben-David, S., Blitzer, J., Crammer, K., Pereira, F.: Analysis of representations for domain adaptation. Adv. Neural Inform. Proc. Syst. **19** (2006)
4. Bengio, Y., Louradour, J., Collobert, R., Weston, J.: Curriculum learning. In: Proceedings of the 26th Annual International Conference on Machine Learning, pp. 41–48 (2009)
5. Bommasani, R., et al.: On the opportunities and risks of foundation models. arXiv preprint arXiv:2108.07258 (2021)
6. Chollet, F.: On the measure of intelligence. arXiv preprint arXiv:1911.01547 (2019)
7. Cody, T.: Mesarovician abstract learning systems. In: Goertzel, B., Iklé, M., Potapov, A. (eds.) AGI 2021. LNCS (LNAI), vol. 13154, pp. 55–64. Springer, Cham (2022). https://doi.org/10.1007/978-3-030-93758-4_7
8. Cody, T.: Homomorphisms between transfer, multi-task, and meta-learning systems. In: Goertzel, B., Iklé, M., Potapov, A., Ponomaryov, D. (eds.) AGI 2022. LNCS (LNAI), pp. 199–208. Springer, Cham (2023). https://doi.org/10.1007/978-3-031-19907-3_19
9. Cody, T., Adams, S., Beling, P.: Motivating a systems theory of AI. Insight **23**(1), 37–40 (2020)
10. Cody, T., Adams, S., Beling, P.A.: A systems theoretic perspective on transfer learning. In: 2019 IEEE International Systems Conference (SysCon), pp. 1–7. IEEE (2019)
11. Cody, T., Beling, P.A.: A systems theory of transfer learning. IEEE Syst. J. **17**(1), 26–37 (2023)
12. De Nicola, R.: Process Algebras. In: Padua, D. (eds) Encyclopedia of Parallel Computing. Springer, Boston, MA (2011). https://doi.org/10.1007/978-0-387-09766-4_450
13. Gilmore, S., Hillston, J.: The PEPA workbench: a tool to support a process algebra-based approach to performance modelling. In: Haring, G., Kotsis, G. (eds.) TOOLS 1994. LNCS, vol. 794, pp. 353–368. Springer, Heidelberg (1994). https://doi.org/10.1007/3-540-58021-2_20
14. Meredith, L.G., Radestock, M.: A reflective higher-order calculus. Electron. Notes Theoret. Comput. Sci. **141**(5), 49–67 (2005)
15. Mesarovic, M.D., Takahara, Y. (eds.): Abstract Systems Theory. Springer, Berlin, Heidelberg (1989)
16. Mitchell, T., et al.: Never-ending learning. Commun. ACM **61**(5), 103–115 (2018)

17. Nivel, E., et al.: Bounded seed-AGI. In: Goertzel, B., Orseau, L., Snaider, J. (eds.) AGI 2014. LNCS (LNAI), vol. 8598, pp. 85–96. Springer, Cham (2014). https://doi.org/10.1007/978-3-319-09274-4_9
18. Pan, S.J., Yang, Q.: A survey on transfer learning. IEEE Trans. Knowl. Data Eng. **22**(10), 1345–1359 (2009)
19. du Preez, A., Beling, P., Cody, T.: A systems theoretic approach to online machine learning. In: 2024 IEEE International Systems Conference (SysCon), pp. 1–8. IEEE (2024)
20. Thórisson, K.R.: A new constructivist AI: from manual methods to self-constructive systems. In: Wang, P., Goertzel, B. (eds.) Theoretical Foundations of Artificial General Intelligence, pp. 145–171. Atlantis Press, Paris (2012). https://doi.org/10.2991/978-94-91216-62-6_9
21. Thórisson, K.R.: Seed-programmed autonomous general learning. In: International Workshop on Self-Supervised Learning, pp. 32–61. PMLR (2020)
22. Thórisson, K.R., Bieger, J., Thorarensen, T., Sigurðardóttir, J.S., Steunebrink, B.R.: Why artificial intelligence needs a task theory. In: Steunebrink, B., Wang, P., Goertzel, B. (eds.) AGI -2016. LNCS (LNAI), vol. 9782, pp. 118–128. Springer, Cham (2016). https://doi.org/10.1007/978-3-319-41649-6_12
23. Wang, Y.: Using process algebra to describe human and software behaviors. Brain Mind **4**, 199–213 (2003)
24. Wang, Y.: Deductive semantics of RTPA. IJCINI **2**(2), 95–121 (2008)
25. Weinshall, D., Cohen, G., Amir, D.: Curriculum learning by transfer learning: theory and experiments with deep networks. In: International Conference on Machine Learning, pp. 5238–5246. PMLR (2018)
26. Weiss, K., Khoshgoftaar, T.M., Wang, D.: A survey of transfer learning. J. Big Data **3**, 1–40 (2016)

Autonomous Intelligent Reinforcement Inferred Symbolism

Berick Cook[1(✉)] and Patrick Hammer[2]

[1] SingularityNET, Palmer, USA
berickcook@gmail.com
[2] KTH Royal Institute of Technology, Stockholm, Sweden

Abstract. This paper introduces AIRIS (Autonomous Intelligent Reinforcement Inferred Symbolism) to enable causality-based artificial intelligent agents. The system builds sets of causal rules from observations of changes in its environment which are typically caused by the actions of the agent. These rules are similar in format to rules in expert systems, however rather than being human-written, they are learned entirely by the agent itself as it keeps interacting with the environment.

Keywords: Procedure Learning · Experiential Learning · Autonomous Agent · Causal Reasoning · Artificial General Intelligence

1 Introduction

With expert systems [3] it is possible to encode simple sets of rules which can represent causal rules inherent in the environment or application domain the AI system is utilized in. These rules govern how the system (or agents of the system) perceive the environment, plan actions, and perform tasks. The rules are hand-coded by human programmers who dictate the behaviors of the system. In environments such as video games, these rules can result in intelligent agents that are capable of sophisticated interactions with their environment, other agents, and the player(s). Unfortunately, these rules are brittle and prone to failing when situations arise that are outside the rules provided by the programmer.

AIRIS is a method for autonomously learning expert system-like casual rules from raw observations of its operating environment. Each rule describes partial changes of state (rather than full state transitions) and together can be used to generate future predicted states via a voting mechanism that takes rule precondition mismatch into account. On a high level this leads to an action state graph which represents an internal world model. However, compared to state transition models as in Markov Decision Processes (MDP, [8]) the agent can construct future states which have not previously been experienced by the agent. The system then uses any planning algorithm on this world model to make plans and perform tasks, while giving it the flexibility to deal with non-stationary environments and changing objectives beyond what reinforcement learning agents [5] with the typical MDP formalization can achieve.

K. R. Thórisson et al. (Eds.): AGI 2024, LNAI 14951, pp. 53–62, 2024.
https://doi.org/10.1007/978-3-031-65572-2_6

AIRIS retains all of the benefits of expert systems such as transparency, mutability, scrutability, and effectivity while also providing the flexibility and adaptability that expert systems lack. AIRIS is continuously learning and not limited to its training. When it encounters situations that are outside the rules it has been trained for, it learns new rules that it can then apply to similar situations. Differently to most reasoning systems, it supports rule induction, which happens while the system is operating and interacting with the environment. Also the system does not need prior logical knowledge to be given, which also sets it apart from projects like Cyc [4]. Additionally, it does not demand logical encodings of inputs, such as as needed by Non-Axiomatic Reasoning System [10]) and planning systems using Planning Domain Definition Language (PDDL, [1]). Instead, it can directly build hypotheses describing action-dependent changes of observable values. AIRIS can be considered an Experiential Learning System which uses an empirical form of causality, which is based on correlation but involving the actions of the agent. This is fundamentally different to Pearl's approach [9] which needs a causal model (a causal graph) to be pre-specified, instead of the agent seeking for (and testing) potentially causal rules through interactions with the environment.

2 Reinforcement Inferred Causality

Expert system rules are comprised of a multitude of conditional statements. The format of these rules is typically IF $< conditions >$ THEN $< statements >$ or a form of logical implication statements [4]. The conditions of these rules are boolean evaluated to determine if the statement should be executed or leads to the described outcome.

AIRIS[1] uses similar conditional statements, referred to as Rules, that it generates through observations. These Rules are used to predict part of a future state by applying the Rules to the current State and storing the newly modified State in a State Graph. The State Graph represents the system's internal world model, and is used to make plans and achieve goals.

Rule Creation: When the agent performs an action, it compares the state of its inputs to the predicted State stored in the State Graph. For each value that is observed to change unexpectedly, a Rule is generated specifically for that value. Therefore the number of Rules that are generated when a change occurs is equivalent to the number of unexpected values that changed. The Conditions for a Rule consist of the action that was taken, the Original Value which is what the value was prior to the action taken, the Relative Change Positions which are the positions relative to the Original Value's position whose values also changed, and the Relative Change Original Values which are what the values in those relative positions were prior to the action taken. The Statement is what the Original Value changed into. When a new Rule is created, the agent's existing

[1] The current implementation is available in the corresponding open-source repository on Github: https://github.com/berickcook/AIRIS_Public.

plan is abandoned, and the State Graph is cleared. A new State Graph is then generated that incorporates the new Rules.

State Graph Creation: A set of Focus Values are compiled from the values in the current State that changed from the previous State. These Focus Values reduce the computational complexity by narrowing the agents attention from every value in the State to a small set of values. It evaluates every Rule for every Focus Value in the State. It computes a precondition match score to evaluate how similar each condition of a Rule is to the current state $(0:1)$. The precondition match score is calculated by comparing the Rule's Condition data with the values in the current State. These precondition match scores are stored in a priority queue. The highest scoring Rule has its Statement applied to the Focus Value. The average of all the highest scoring Rules is the State Confidence $(0:1)$ of the prediction. The values in the positions in the highest scoring Rule's Relative Change Positions are added to the Focus Values. This ensures that other relevant values in the State are evaluated if they were not part of the original Focus Values. The resulting modified State is stored in the State Graph with an Edge labeled with the action taken and State Confidence of the prediction connecting it back to the original State. This results in a self-generated State Graph that mirrors the style and usefulness of finite-state machines. And since it is self-generated the possible states are not pre-specified, they result from the application of the causal rules which capture changes of individual values rather than the complete state. This fosters the generalization abilities of AIRIS as it allows the system to use its knowledge in novel situations it has never seen before.

State Graph Usage: The State Graph initially consists of the current state of the environment inputs, and any number of unconnected Goal Sub-States. Goal Sub-States are generated sub-states that consist solely of the conditions of Rules where the Statement results in achieving the agent's specified Goal. If the agent has not yet observed a change that resulted in achieving the agent's Goal, then the State Graph only consists of the current environment State. If such Goal Sub-States exist, then as each new state is added to the State Graph it is heuristically evaluated to determine how close or far it is from all Goal Sub-States. This heuristic guides the Planning process to generate new States towards the closest unconnected Goal Sub-State. When a new state is generated that contains the Goal Sub-State, a Goal State has been found. The action path from the current State to the Goal State becomes the agent's plan. If no Goal Sub-States exist or the Goal Sub-States cannot be reached, then the State with the lowest Confidence becomes the Goal State. This results in the agent pursuing novel situations to observe and generate new Rules. The new Rules may allow it to create a Goal Sub-State or predict a new path to an existing Goal Sub-State. This can be considered as an effective model of artificial curiosity, which does not only capture aspects of information gain as in [7], but leads to deliberate interactions planned by the agent using the empirically-causal knowledge it possesses.

3 Architecture

The architecture of AIRIS consists of four primary components (Fig. 1). The Rule Base of learned rules, the State Graph that contains the world model of current and future predicted states, the Pre-Action Observation sequence, and the Post-Action Observation sequence.

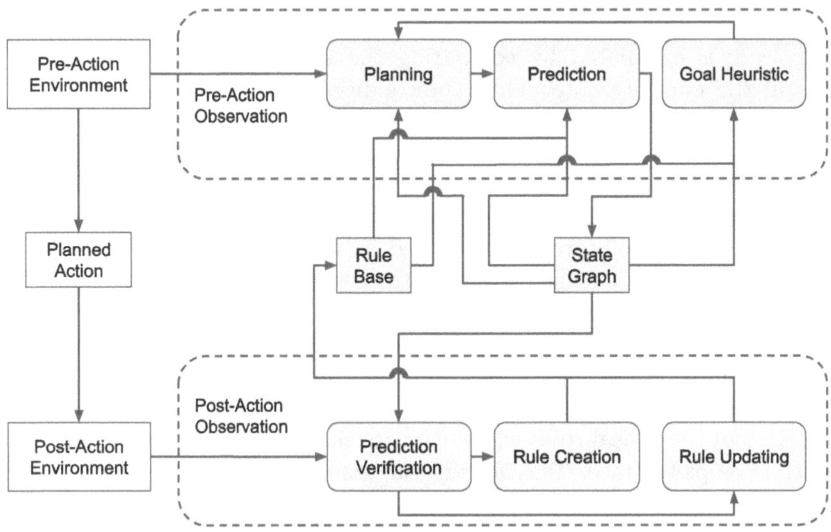

Fig. 1. AIRIS architecture diagram

Pre-Action Environment: The collection of sensory inputs from the operating environment prior to an action being taken.

It consist of sub-symbolic values which capture the presence of different object types at the existing locations of the environment.

Pre-Action Observation: This is the sequence of processes to make predictions of states to store in the State Graph and search across the State Graph for sequences of actions that result in achieving the current Goal.

Planning: This process guides the generation of new prediction States towards Goal Sub-States. If a Goal State is found, it traces the sequence of actions necessary to traverse the State Graph from the current State to the Goal State and sequentially feeds the actions back to the environment. If there are no Goal Sub-States or Goal Sub-States cannot be reached then the State with the lowest State Confidence becomes the Goal State.

Prediction: This process takes a given state from the Planning process and predicts what the next state will be for each action available based on the rules in the Rule Base. These predicted states are then added to the State Graph.

Goal Heuristic: This is a dynamically calculated heuristic value that is assigned to each state in the State Graph. It takes every rule from the Rule Base whose Statement results in achieving its current goal, and compares it to a given state from the State Graph. If the values in the conditions of the rule exist in the state data, it calculates the relative Manhattan distance between those values and assigns the distance as the heuristic value for that state. A State's heuristic value is used by the Planning process to guide predictions toward the Goal.

Post-Action Environment: The collection of sensory inputs from the operating environment after an action being taken.

Post-Action Observation: This is the sequence of processes to analyze observations of changes that occurred when the Planned Action was performed, and to create new rules or update the data in existing rules.

Prediction Verification: This process compares the Post-Action Environment data to the predicted State data to verify that a prediction was correct. If there is any discrepancy, then the action plan is abandoned, the State Graph is cleared, and new rules are created for the failed predictions. If there is no discrepancy, then the rules that the prediction was based on are updated with any relevant information.

Rule Creation: If the Prediction Verification process finds a discrepancy, then the Rule Creation process builds new rules that represent the unexpected changes. These rules are then added to the Rule Base.

Rule Updating: If the Prediction Verification process does not find a discrepancy, the Rule Updating process updates the rules that were applied in the accurate prediction.

4 Data Structures

There are two primary classes of data: Rules and States. Rules consist of any number of Conditions which were observed to be true when a value in the environment changed, and what the value changed into. Conditions consist of the relative position of other values in the environment that also changed, what those values changed into, and a list of other rules that were observed to be true for the prediction. States consist of copies of the environment data that have been modified by the Prediction process and represent future predicted states, a list of the Rules that were applied during the Prediction process, the action that was performed that resulted in the State, and a reference to the prior State.

5 Experiments and Comparisons

AIRIS has been tested in several different environments to examine its general purpose capabilities.

Grid World Puzzle Game: The primary testing environment has been a simple grid world puzzle game. The purpose of this environment is to test its ability to learn simple game rules, use those rules to solve problems and complete tasks with causal reasoning, demonstrate the ability to change its objectives with no impact on efficacy, and demonstrate the overall transparency and scrutability (Fig. 2).

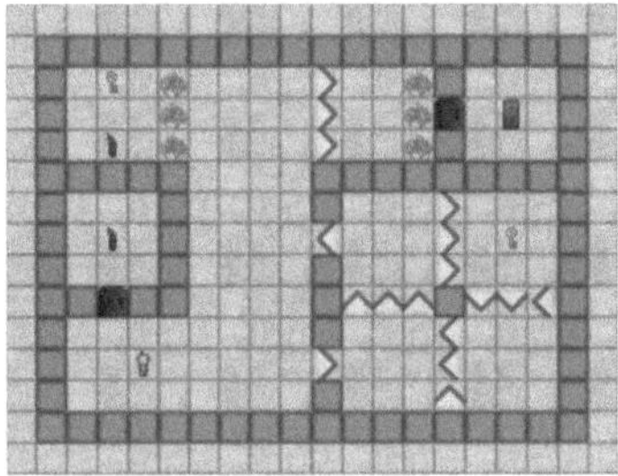

Fig. 2. Level 12 of the Grid World Puzzle Game

Inputs consist of a two dimensional array of integers that represent the objects in the game world, and a one dimensional array of integers that represent the inventory of items collected by the agent. Outputs consist of four actions: up, left, down, and right. The game consists of a series of thirteen levels of varying difficulty. The game objects consist of the agent, walls, batteries, one-way arrows, doors, keys, fire, and fire extinguishers. The agent completes a level by collecting all of the batteries. One-way arrows cannot be moved into from the direction they are pointing, but can be moved into from any of the other three directions. Doors require a key to open which consumes the key. Fire restarts the level on touch and subtracts one battery, but can be put out by a fire extinguisher on touch which consumes the fire extinguisher. There is no time limit.

The untrained agent began at level one and played sequentially through the levels in one continuous episode. If the agent touched fire and restarted a level, the episode did not end. It was only given the Goal of maximizing the number of batteries collected. The agent first discovered how to collect a battery after 32 timesteps. The agent was then able to complete all levels in the first episode in 4,612 timesteps. This demonstrates that the agent is able to adapt to novel situations such as encountering new obstacles and levels without the need for extensive training over multiple episodes.

The trained agent was placed back at level one and it was able to complete all levels in the second episode in 921 timesteps. This demonstrates that the agent is able to utilize the rules it learned to rapidly improve its efficacy.

The trained agent was placed into a new level that it had not been trained on that contained all game objects. It was able to solve it in 67 timesteps (Best possible is 54 timesteps). This demonstrates that the agent is not limited to levels it has been trained on, and can apply causal reasoning to achieve goals in novel level configurations (Fig. 3).

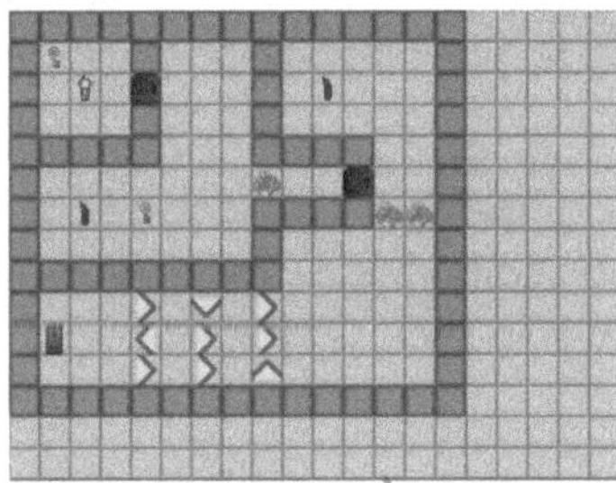

Fig. 3. AIRIS can complete a level it has never seen

The trained agent was placed in level eight. Before it could collect the final battery, its Goal was changed to decrease batteries collected. It immediately abandoned collecting the battery and went into the fire. When the level reset, it ignored the doors, keys, and batteries and went into the fire again. It continued to decrease its battery count until the Goal was changed back. This demonstrates that the Goal of the agent can be changed at any time.

The trained agent was placed back at level one with the Goal of decreasing its battery count. It is not possible for the agent to decrease its battery count until it encounters fire in level seven. The agent took slightly longer than usual to complete the levels, as it explored predictions with less than 100% confidence. It only collected batteries when there was nothing more for it to explore. When it reached level seven it immediately and repeatedly touched the fire. This demonstrates that even when its Goal is not possible the agent will explore until it can achieve its Goal even if the Goal was not what it was originally trained on.

The trained agent was placed back at level one and a visualization tool was used to view the agent's internal world model and see what its next plan is. When the visualization tool is active, the agent waits for a command to proceed before executing its plan. This demonstrates the scrutability of the system as are able to directly interface with the agent's internal world model to see what it has learned, what its predictions are, and what it is planning to do. The visualization tool can be expanded upon to show any additional details that the user may find relevant, or allow the user to control and "play" the internal world model as though the user was playing the game itself (as demonstrated by NACE).

Cartpole: This classic control problem which is concerned to balance a pole on a cart [2], tested the ability of AIRIS to work in a continuous space for potential future robotic applications, as well as tested it against a well established Reinforcement Learning benchmark.

Inputs consist of a one dimensional, floating-point array containing the Cart Position, Cart Velocity, Pole Angle, and Pole Angular Velocity. Outputs consist of two actions: push cart left and push cart right. The Goal is to keep the Pole Angle as close to 0 as possible for 200 time steps. If the Pole Angle exceeds $\pm 12°$ or the Cart Position is greater than ± 2.4 then the episode ends. Achieving a score of 195 over the last 100 episodes is considered solved. The environment also provides a reward, but it was not utilized as the goal condition is considered in planning instead.

AIRIS was able to reach its first score of 200 after 32 episodes and solved the environment after 130 episodes with consistent success. Comparatively, DQN took over 60 episodes to reach its first score of 200, and solved the environment after 200 episodes with varying success (Fig. 4).

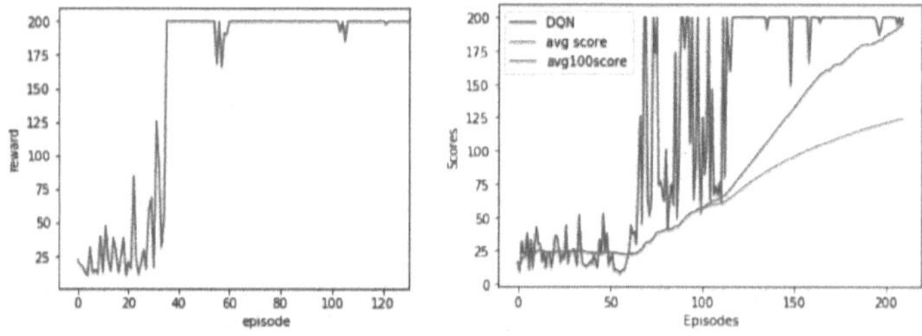

Fig. 4. AIRIS Cartpole results (left) and DQN Cartpole results (right)

MNIST: This classic image recognition problem is concerned about recognition of handwritten digits. With this dataset we tested the potential multi-modal capabilities of AIRIS. While Convolutional Neural Networks represent the state-of-the-art in this dataset in terms of accuracy after training, rule-based learning like in AIRIS can lead to data-efficient learning and hence also deserves to be investigated.

Inputs consist of a two dimensional array of pixel data from a "Training" set of 60,000 handwritten numbers and a "Test" set of another 10,000 numbers, and a single value one dimensional array of the label of the current image. Output consists of a single action "Reveal Label". An image is shown to the system with a null label. The system must then predict what value the label will be when the action is performed.

On the first run of the "Training" set, AIRIS's accuracy quickly rose and finished with 91% accuracy. The trained agent was then given the "Test" set where its accuracy dropped to a low of 90% before finishing at 93%. On the second run of the "Training" set, it had a consistent accuracy of 97%. On the second run of the "Test" set it had a consistent accuracy of 99%. Comparatively, the deep Convolutional Neural Network [6] took several hundred runs to reach equivalent accuracy (Figs. 5 and 6).

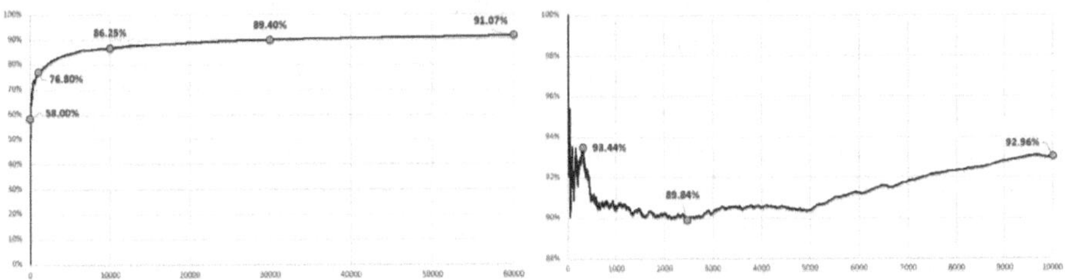

Fig. 5. AIRIS MNIST training set (left) and test set (right) results from first run

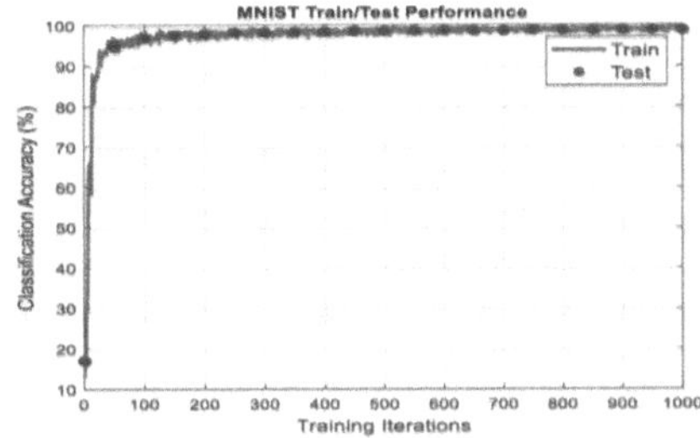

Fig. 6. Optimized Convolutional Neural Network Model Results

6 Limitations

As of this paper, AIRIS is not able to learn rules for events that occur regardless of the agent's action. Such as the independent movements of other agents or objects. This limits the system to environments where the only events that occur are caused directly by the actions of the agent. Environments where all input values can change, such as video feeds, pose a computational cost limitation. AIRIS will likely require additional attention mechanisms to prioritize processing of values relevant to the current task and disregarding other values to optimize the computational cost. Additionally, while an extension to three dimensions is straightforward, more research is needed for continuous spaces or high-resolution semantic grid maps. The latter will be required for letting AIRIS operate autonomous mobile robots, as grid maps provide a suitable representation to make AIRIS well-integrated with Robot Operating System. We expect that this will allow the system to operate in real-world scenarios when variants of Semantic Simultaneous Localization and Mapping are utilized.

7 Conclusion

Despite its current limitations, AIRIS is a promising alternative to contemporary Reinforcement Learning techniques. In grid world environments it can learn from a fraction of the training that RL techniques require, with substantially better efficacy due to its causal representations, and has the ability to immediately switch to different objectives. Its ability to continuously learn allows it to adapt to new situations outside of its training, and its transparency and scrutability offers many advantages over RL. This system can help to increase the generality of autonomous agents and the data-efficiency of learning goal-directed behaviors.

Future work will address the aforementioned limitations: extending the model to deal with agent-external changes, partial observability, and non-determism, as well as to extend the model to work in continuous (or high-resolution) three-dimensional environments. Also, we will more systematically analyze the inductive biases which make AIRIS so data-efficient.

References

1. Aeronautiques, C., et al.: PDDL-the planning domain definition language, Technical report, Yale Center for Computational Vision and Control (1998)
2. Geva, S., Sitte, J.: A cartpole experiment benchmark for trainable controllers. IEEE Control Syst. Mag. **13**(5), 40–51 (1993)
3. Giarratano, J.C., Riley, G.: Expert Systems. PWS Publishing Co., United States(1998)
4. Lenat, D.B., Guha, R.V., Pittman, K., Pratt, D., Shepherd, M.: Cyc: toward programs with common sense. Commun. ACM **33**(8), 30–49 (1990). https://doi.org/10.1145/79173.7917
5. Moerland, T.M., Broekens, J., Plaat, A., Jonker, C.M., et al.: Model-based reinforcement learning: a survey. Found. Trends Mach. Learn. **16**(1), 1–118 (2023)
6. Salman, O.S., Salman, A.S.: Addressing challenging problems using optimized deep learning classification algorithms on the MNIST dataset. In: Arai, K. (ed.) Advances in Information and Communication: Proceedings of the 2022 Future of Information and Communication Conference (FICC), Volume 2, pp. 247–260. Springer, Cham (2022). https://doi.org/10.1007/978-3-030-98015-3_17
7. Still, S., Precup, D.: An information-theoretic approach to curiosity-driven reinforcement learning. Theory Biosci. **131**, 139–148 (2012)
8. Sutton, R.S., Barto, A.G.: Reinforcement Learning: An Introduction. MIT press (2018)
9. Verma, T., Pearl, J.: Causal networks: semantics and expressiveness. In: Machine Intelligence and Pattern Recognition, vol. 9, pp. 69–76. Elsevier (1990)
10. Wang, P.: Non-Axiomatic Logic: A Model of Intelligent Reasoning. World Scientific (2013)

Decoding Chess Mastery: A Mechanistic Analysis of a Chess Language Transformer Model

Austin L. Davis and Gita Sukthankar(✉)

University of Central Florida, Orlando, FL, USA
{austindavis,gitasukthankar}@ucf.edu

Abstract. Mechanistic interpretability (MI) studies aim to identify the specific neural pathways that underlie decision-making in neural networks. Here we analyze both the horizontal and vertical information flows of a chess-playing transformer. This paper introduces a new taxonomy of chessboard attention patterns that synchronize to guide move selection. Our findings show that the early layers of the chess transformer correctly identify moves that are highly ranked by the final layer. Experiments conducted on human chess players laid the foundation for much of our current understanding of human problem-solving, cognition, and visual memory. We believe that the study of chess language transformers may be an equally fruitful research area for AGI systems.

Keywords: chess cognition · mechanistic interpretability · transformers

1 Introduction

This paper presents a mechanistic analysis of the operation of a chess language transformer model. In contrast to other surface-level forms of interpretability that simply correlate inputs with outputs or highlight which features were important for a decision [13], the mechanistic interpretability community [14] seeks to uncover the computational processes that occur within a neural network, literally reverse-engineering the algorithms to reveal how the model arrived at its conclusions. This approach moves beyond treating the model as an opaque "black box" and instead examines the model's choices through the lens of its internal operations. Reverse-engineering transformer models has become an important area of research interest for two reasons:

1. transformers are a key building block for many of the SOTA vision, language, and multimodal visual-language systems (e.g., ViT [7]), ChatGPT, Gemini, and CLIP [17]);
2. the architectural elegance of the transformer makes it more amenable to certain types of circuit analysis. Elhage et al. [8] note that if MLP layers are removed, the residual stream of a transformer is composed of linear and additive operations, in which the attention heads operate independently.

K. R. Thórisson et al. (Eds.): AGI 2024, LNAI 14951, pp. 63–72, 2024.
https://doi.org/10.1007/978-3-031-65572-2_7

This paper specifically examines the operation of the attention heads in a chess language transformer and proposes a taxonomy of chess attention mechanisms. Although there are several chess language transformer models, we selected Toshniwal et al.'s Learning Chess Blindfold (LCB) model [20] for our analysis, due to its manageable size. LCB uses the GPT-2 small architecture [18] and is trained on sequences from human chess games. This paper explores the distribution of attention patterns per layer. Our results show that early layers of the LCB are very successful at predicting moves that are ultimately highly ranked by the final layer.

2 Related Work

2.1 Chess and Cognition

There is a long tradition of using chess as a benchmark for artificial intelligence systems ranging from specialized parallel search algorithms like Deep Blue [3] to more generalized machine learning approaches such as AlphaZero [11]. However, our work is inspired by research from the cognitive science community that endeavors to model chess mastery in humans. Charness [4] quantified the impact of chess on cognitive science by doing a literature review of published studies referencing the seminal article by Simon and Chase [5] that proposed the usage of chess for cognitive science research. Neural imaging studies [9] on human chess players have shown that expert chess players exhibit some differences in grey matter within the occipitotemporal junction (OTJ), a region linked to various perceptual processes. FMRI studies of chess [2] show increased activity in brain regions associated with spatial reasoning (parietal areas) with less activity in the frontal lobe. These findings suggest that spatial processing may play a more prominent role than decision-making in human chess cognition. Our study of chess attention patterns in transformers parallels these human fMRI studies. However most of the human fMRI studies are conducted by simply showing the subjects images of chess boards, whereas we examine average attention patterns across many games.

2.2 Mechanistic Interpretability (MI)

In this paper, we apply several techniques from the MI community to a chess-playing language model. Mechanistic interpretability is a newer area of explainable AI in which the aim is to reverse engineer neural circuits [14]. This can be accomplished through a variety of techniques including linear probes, ablation studies, and activation patching [19]. However, all these methods rely heavily on human inspection. To ameliorate this limitation, Conmy et al. [6] recently proposed a method for automating the circuit detection step that successfully rediscovered circuits in GPT-2, identified by prior work.

Many researchers believe that large language models obey the *linear representation hypothesis* [16]. This theory suggests important concepts that the model learns are encoded in a linear way within its hidden layers, facilitating the probe training process. Linear probes have been used in several of the prior studies on game play language models to determine how the transformer encodes game state [10,15,20]. In this paper, we ignore the question of world state and simply examine the horizontal and vertical information flows.

3 Method

3.1 The Learning Chess Blindfolded (LCB) Model

Toshniwal et al.'s Learning Chess Blindfolded (LCB) model [20] utilizes the GPT-2 small architecture [18] with 12 attention heads in each of its 12 transformer layers. LCB is a causal language model trained to minimize loss in predicting the next token in chess games. Its inputs are sequences of chess moves (game traces) given in Universal Chess Interface (UCI) notation. UCI notation describes chess moves by specifying the starting and ending square of a piece's movement. For example, the move of a knight from b1 to c3 is noted as "b1c3". Figure 1 shows how a game trace in UCI notation correspond to the chess board state.

Partial game trace in UCI notation:

`e2e4 c7c5 g1f3 d7d6 d2d4 c5d4 f3d4 g8f6 b1c3 a7a6 c1e3 e7e5 d4b3 c8e6 f2f3 f8e7 d1d2 e8g8 e1c1 a6a5 f1b5 b8a6 c1b1 a6c7 ...`

Board state after each complete move:

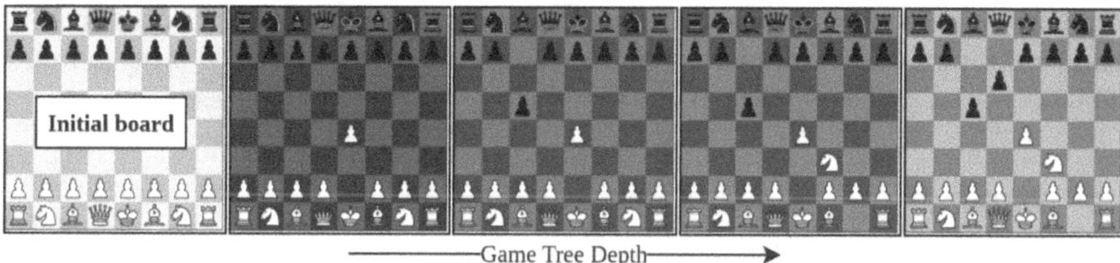

Fig. 1. UCI notation of a partial game trace and the resulting board states after each move. The game starts from the initial board position, and the first move shown in red moves the white pawn from e2 to e4. LCB treats the start and end tiles of a single move as separate tokens. Thus, each move requires two tokens. (Color figure online)

To input a game trace into the LCB model, each UCI move is divided into two tokens: one for the starting tile and another for the destination tile, each processed separately. The model outputs a non-normalized vector over its vocabulary (see Table 1), which is converted into a probability distribution using the `softmax` operation. Predicting a chess move involves inputting a game trace prefix and prompting the model twice. In the first forward pass, the starting tile is selected from the output probability distribution, appended to the trace, and the model is prompted again to determine the destination tile. This interactive process is facilitated through a Google Colab notebook[1].

[1] See colab.research.google.com/drive/125y4MpnSWAakoSE5My9jGMtBExbjqTpW.

Table 1. LCB Vocabulary

Type	Examples	Count
Square names	`e4`, `d1`	64
Promotion type	`q`, `r`, `b`, `n`	4
Special symbols	BOS, EOS, PAD	9
Total		77

LCB predicts legal moves with an accuracy of 97.7% [20]. In practice, we find it can regularly play thirty complete moves before it recommends an illegal move. This is impressive when considering that the model was trained with no *a priori* knowledge of the game of chess; it was trained exclusively to minimize loss when predicting the next token in UCI move sequences. Consequentially, LCB does not play to win; it simply anticipates which move is most likely to occur next. As a result, LCB plays stronger moves if the input sequence resembles high-skill gameplay and weaker moves when the input sequence resembles low-skill gameplay.

Our intent was to localize the internal mechanisms which allow it to generate legal moves with such high accuracy. Importantly, our use of the LCB model is strictly out-of-the-box, meaning we have not modified its configuration or parameterization. In a similar way that a neural scientist might localize a cognitive function by stimulating a patient with sound or images during an MRI scan, we stimulate the model by performing forward passes on a large number of chess games and record the values of the model's hidden states.

3.2 Dataset

We utilize a dataset of 1000 chess games sourced from the lichess.org open database. The games were converted into UCI notation and deduplicated. Each game in the dataset is truncated to the first 20 tokens (representing the first 10 moves) to center analysis on opening move sequences, provide consistency when computing metrics, and focus on games without pawn promotions since promotion is handled differently by the model and tokenizer.

3.3 Activation Caching and Direct Logit Attribution (DLA)

During each forward pass, the hidden state $h_i^{(\ell)}$ is captured post each transformer layer in an *activation cache*. This activation cache is the set

$$\text{ActivationCache} = \{h_i^{(\ell)} : i \in [1, 12] \text{and} \ell \in [1, \texttt{numTokens}]\}$$

We employ DLA [1] to assess the impact of each layer on the decision-making process of LCB. By applying the GPT's `unembed` matrix directly to the captured hidden states, we can back-translate these intermediate representations from the

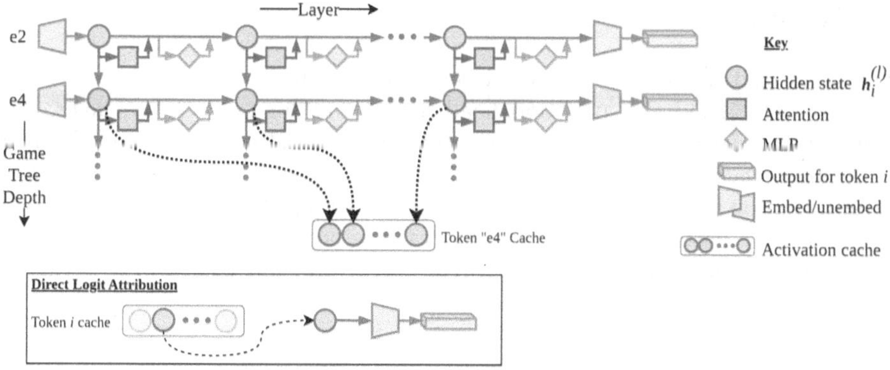

Fig. 2. Caching hidden states within the GPT-2 architecture. DLA applies the `unembed` matrix to a vector in the activation cache to transform latent hidden state vectors into a vector in the vocabulary (output) space. By comparing DLA vectors across layers we can see how each layer contributes to the final output of the LCB model. The unfolded representation of the GPT-2 architecture was inspired by [12].

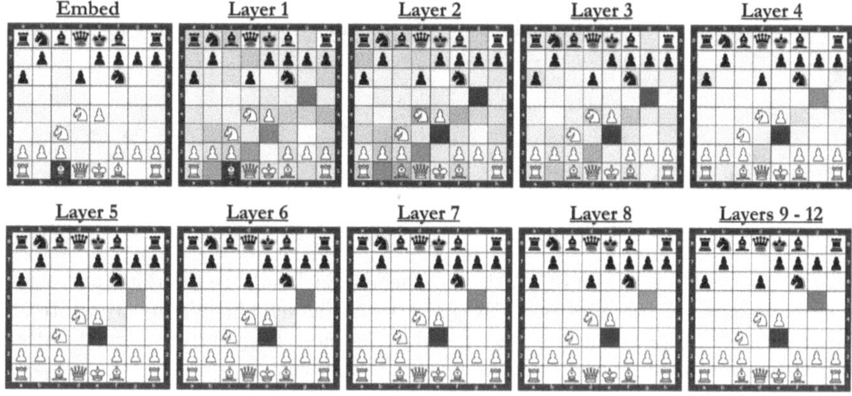

Fig. 3. DLA visualization. The model selects the destination square for the bishop on `c1`. Boards are colored according to the softmax of the unembedded hidden state for each layer starting from the token embedding (top-left) through layer 12 (bottom-right); richer blue colors indicate stronger activations. Layer 1 activations highlight the bishop on c1 (the piece about to be moved), and intermediate layers process information in the previous hidden states to select a destination for that bishop. The model's final (actual) output, `e3`, corresponds to the darkest tile on the bottom-right board. The model's preference for `e3` is evident as early as layer 2. (Color figure online)

latent embedded space to the output vocabulary. By analyzing the evolution of these intermediate states, we can uncover how each layer influences the final move prediction. This process is illustrated in Figs. 2 and 3.

4 Results

We performed a forward pass on 1,000 unique game traces with at least 20 tokens, and cached the attention head activations for every head, layer, and game trace. We averaged these activations over the 1,000 games to obtain a 20×20 activation pattern for all 144 ($= 12 \times 12$) attention heads. Inspecting these attention activations, we observed several, strong structural patterns shown in Fig. 4. Each pattern emerges from the training process and implies these heads are selecting information in a very predictable way that corresponds with the regular structure of UCI game traces.

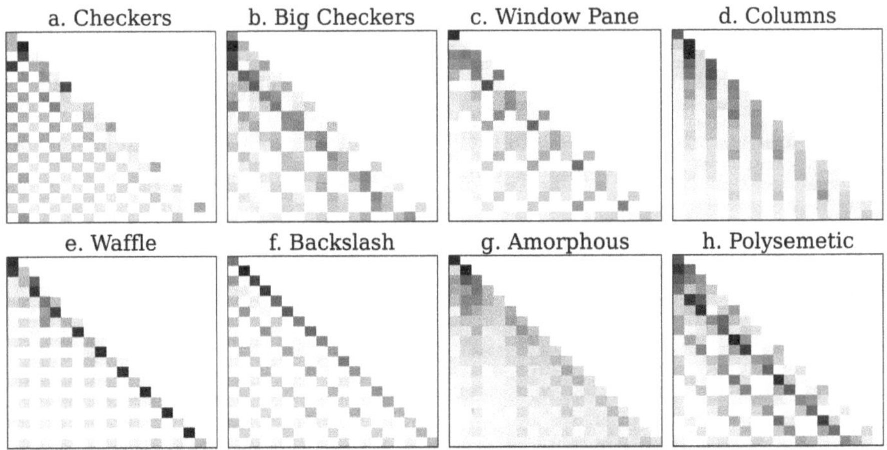

Fig. 4. Structural patterns in LCB attention heads. Each pattern (a–h) is a prototypical example. Values are averages over 1K game traces which range from 0 (white) to 1 (black) with $\sigma = 0.133$. The top row includes the most frequently observed patterns. The bottom row includes additional noteworthy patterns.

We interpret these attention patterns from the perspective of UCI game traces. (Note: a full move is processed in either 2 forward passes (for a single player) or 4 forward passes (to complete a full turn for both white and black players).

The following patterns were observed multiple times in the attention heads of the LCB model. Their frequencies are reported in Table 2:

a. **Checkers**. When selecting a source/destination tiles, these heads attend to the history of previous source/destination tiles. These heads do not appear to have a limited context window, and the full history of moves are typically attended to.
b. **Big Checkers** Similar to the checkers pattern, except rather than attending to the history of tiles, these heads attend to the history of full moves. Depending on the horizontal offset of the grid, these heads either attend to the current player's move history or the opponent's move history.

c. **Window Pane** These heads have two modalities. In the prototypical example above, LCB attends only to white player's destination tile history exactly when selecting the destination tile for white player pieces. During the other phases of movement, the head attends approximately equally to all positions.
d. **Columns** These heads attend to the history of a single phase of movement, i.e. either destination or source tile selection depending on the horizontal offset of the columns.

The following patterns were observed at least once, and were noteworthy because of their strong structural regularity:

e. **Waffle** This pattern occurred once in layer 3 head 5 (L3H5). When selecting a destination, the head attends only to the previous source tile (because of the strong diagonal), and when selecting a source tile, the head attends to both players' previous destinations equally.
f. **Backslash** (L1H5 and L1H12) This is a special case of the checkers pattern, but instead of looking back one phase, these heads look back four tokens to the player's own source/destination history.
g. **Amorphous** (L12 H1-H3) These heads occurred exclusively in the final layer. Unlike other heads which have a high variance, these heads are diffuse, indicating the head is likely highly state-dependent.
h. **Polysemetic** Many heads can be described as a combination of multiple other patterns, encoding multiple modalities at once. The head pictured here (L4H12) is simultaneously a Big Checkers and a Backslash pattern.

Table 2. Frequency Counts by Layer for Top Four Attention Patterns

Layer	Checkers	Big Checkers	Columns	Window Pane
Layer 1				1
Layer 2	1		1	1
Layer 3	1	2	1	
Layer 4	1	3	1	
Layer 5		5	1	1
Layer 6		5		2
Layer 7		3		2
Layer 8				
Layer 9		1	3	
Layer 10		1		
Layer 11				
Layer 12				
Total	3	20	7	7

4.1 Comparing DLA Output Vectors Across Layers

A key question is when the LCB model completes its computation during the forward pass. By comparing the final output vector with a DLA vector from any layer, we can use their similarity as an indicator. A small metric value between layer ℓ and layer 12 suggests that computation is largely complete by layer ℓ.

We employ three metrics to evaluate similarity between DLA vectors and the final output. **Cosine similarity** measures vector alignment, with higher values indicating greater similarity. **Precision at K (P@K)** evaluates how well the top K predictions from earlier layers match the final prediction, at levels $K = 1$, $K = 2$, and $K = 5$. To address gaps left by these metrics, we introduce **Magnitude Precision at K (M-P@K)**, which considers vector magnitudes.

Let S be the scores tensor and T be the target tensor (i.e. the unnormalized output of layer 12). Define T' as the softmax-normalized version of T. Select the top k entries from S and T', denoted as S_k and T'_k respectively. Let idx_k be the indices corresponding to the top k entries from S. The corresponding targets from T' at these indices are $T'[\text{idx}_k]$. Compute the following magnitudes:

$$\text{Mag}_{\text{sel}} = \sum T'[\text{idx}_k], \quad \text{and} \quad \text{Mag}_k = \sum T'_k$$

Then the Magnitude Precision at K (MP@K) is the ratio of these magnitudes: $\text{MP@K} = \text{Mag}_{\text{sel}}/\text{Mag}_k$. Figure 5 and Fig. 6 show the values of these metrics over the 1000 game traces used to perform the attention head analysis, the mean is taken over token positions $i = 1, ...20$.

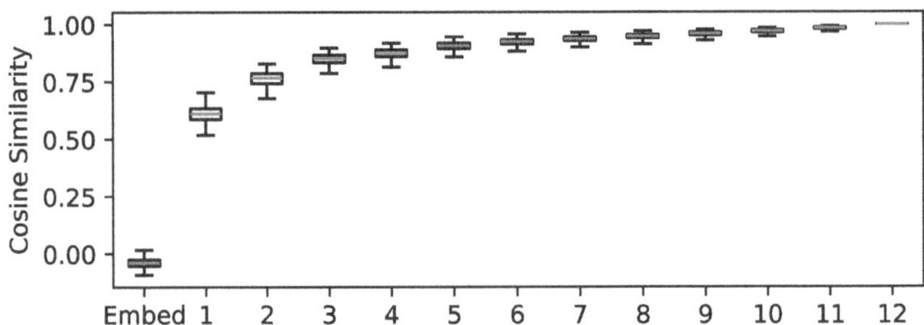

Fig. 5. Cosine similarity for DLA of h_i^ℓ versus output at layer 12. *Embed* is the hidden state vector prior to layer 1, just after `embed()` on the token id.

Figure 5 reveals significant insights. The `embed()` function's vector is nearly perpendicular to the final output vector, aligning with observations from Fig. 3 that the Embed layer indicates the moved piece based on the latest input token `c1`, rather than its destination. By layer $\ell = 9$, much of the computation is already complete as the median cosine similarity reaches 92%.

In interpreting Fig. 6, note the presence of multiple viable chess moves per position, affecting the stability of rank-based metrics like P@K and M-P@K. The

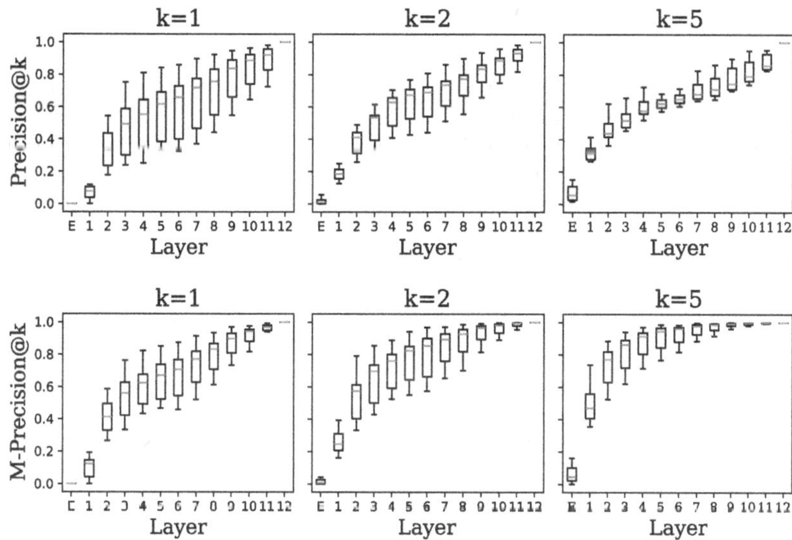

Fig. 6. Precision at k and Magnitude (M-)Precision at k by layer. E = *Embed*

large inner quartile range of $P@1$ across all layers, which narrows significantly for $k > 1$, suggests that while the model often identifies several plausible moves early, the optimal move typically emerges in the last three layers. Additionally, the relative stability of the M-P@K metric from $k = 1$ to $k = 2$ indicates that even if the top-ranked move at earlier layers differs from the final layer, the quality of alternative choices remains high.

5 Conclusion

This paper presents a mechanistic analysis of the LCB chess-playing transformer. We introduce a new taxonomy of chessboard attention patterns and describe how they can be used to guide chess play. The first seven layers of the transformer use our attention patterns frequently. According to our ranked move analysis, these are also the layers that contribute the most to move selection. High metric values for layers $\ell \geq 8$, combined with the distribution of attention head patterns in Table 2 and manual reviews of numerous DLA sequences such as those shown in Fig. 3, suggest a significant interaction between structured attention heads and move selection in the LCB model. However, the function of the final five layers remains inconclusive. They likely play a role in long-term strategy, which is difficult to analyze using current techniques.

Acknowledgments. The conclusions and opinions expressed in this research paper are those of the authors and do not necessarily reflect the official policy or position of the U.S. Government or Department of Defense.

References

1. Interpreting GPT: The Logit Lens. https://www.lesswrong.com/posts/interpreting-gpt-the-logit-lens
2. Atherton, M., Zhuang, J., Bart, W.M., Hu, X., He, S.: A functional MRI study of high-level cognition. I. the game of chess. Cogn. Brain Res. **16**(1), 26–31 (2003)
3. Campbell, M., Hoane Jr, A.J., Hsu, F.H.: Deep blue. Artif. Intell. **134**(1-2), 57–83 (2002)
4. Charness, N.: The impact of chess research on cognitive science. Psychol. Res. **54**, 4–9 (1992)
5. Chase, W.G., Simon, H.A.: The mind's eye in chess. In: Visual Information Processing, pp. 215–281. Elsevier (1973)
6. Conmy, A., Mavor-Parker, A., Lynch, A., Heimersheim, S., Garriga-Alonso, A.: Towards automated circuit discovery for mechanistic interpretability. Adv. Neural. Inf. Process. Syst. **36**, 16318–16352 (2023)
7. Dosovitskiy, A., et al.: An image is worth 16x16 words: transformers for image recognition at scale. arXiv preprint arXiv:2010.11929 (2020)
8. Elhage, N., et al.: A mathematical framework for transformer circuits. Transform. Circuits Thread **1**, 1 (2021)
9. Hänggi, J., Brütsch, K., Siegel, A.M., Jäncke, L.: The architecture of the chess player's brain. Neuropsychologia **62**, 152–162 (2014)
10. Li, K., Hopkins, A.K., Bau, D., Viégas, F., Pfister, H., Wattenberg, M.: Emergent world representations: exploring a sequence model trained on a synthetic task. In: International Conference on Learning Representations (2023)
11. McGrath, T., et al.: Acquisition of chess knowledge in AlphaZero. Proc. Natl. Acad. Sci. **119**(47), e2206625119 (2022)
12. Meng, K., Bau, D., Andonian, A., Belinkov, Y.: Locating and editing factual associations in GPT. In: NIPS, vol. 35, pp. 17359–17372 (2022)
13. Minh, D., Wang, H.X., Li, Y.F., Nguyen, T.N.: Explainable artificial intelligence: a comprehensive review. Artif. Intell. Rev. 1–66 (2022)
14. Nanda, N.: A comprehensive mechanistic interpretability explainer and glossary. https://www.neelnanda.io/mechanistic-interpretability/glossary
15. Nanda, N., Lee, A., Wattenberg, M.: Emergent linear representations in world models of self-supervised sequence models (2023)
16. Park, K., Choe, Y.J., Veitch, V.: The linear representation hypothesis and the geometry of large language models (2023)
17. Radford, A., et al.: Learning transferable visual models from natural language supervision. In: International Conference on Machine Learning, pp. 8748–8763. PMLR (2021)
18. Radford, A., Wu, J., Child, R., Luan, D., Amodei, D., Sutskever, I.: Language models are unsupervised multitask learners (2019)
19. Räuker, T., Ho, A., Casper, S., Hadfield-Menell, D.: Toward transparent AI: a survey on interpreting the inner structures of deep neural networks (2023)
20. Toshniwal, S., Wiseman, S., Livescu, K., Gimpel, K.: Chess as a testbed for language model state tracking. In: Proceedings of the AAAI Conference on Artificial Intelligence, vol. 36, pp. 11385–11393 (2022)

Argument-Driven Planning and Autonomous Explanation Generation

Leonard M. Eberding[1(✉)], Jeff Thompson[2], and Kristinn R. Thórisson[1,2]

[1] Center for Analysis and Design of Intelligent Agents, Reykjavik University, Reykjavik, Iceland
{leonard20,thorisson}@ru.is

[2] Icelandic Institute for Intelligent Machines (IIIM), Reykjavík, Iceland
jeff@iiim.is

Abstract. Research on general machine intelligence is concerned with building machines that are capable of performing a multitude of highly complex tasks in environments as complex as the real world. A system placed in such a world of indefinite possibilities and never-ending novelty must be able to adjust its plans dynamically to adapt to changes in the environment. These adjustments, however, should be based on an informed explanation that describes the hows and whys of interventions necessary to reach a goal. This means that explanations are at the core of planning in a self-explaining way. Using Assumption-Based Argumentation we present a way how an AGI-aspiring system could generate meaningful explanations. These explanations consist of argumentation graphs that represent proponents (i.e., solutions to the task) and opponents (contradictions to these solutions). They thus provide information on why which intervention is necessary, thus making an informed commitment to a particular action possible. Additionally, we show how such argumentation graphs could be used dynamically to adjust plans when contradicting evidence is observed from the environment.

Keywords: Argumentation · Artificial Intelligence · General Machine Intelligence · Causal Reasoning · Self-Explanation

1 Introduction

The process of explanation involves more than a transmission of information between two or more agents. An explanation can help a system, in a self-explaining manner, to identify reasons for and against achieving a goal at a future time, e.g. when deciding what actions to take or what plans to commit to [12]. For this, however, a special type of explanation is necessary. This type of explanation differs from the idea of explanations (or interpretations, two concepts often used interchangeably) in Deep Learning (DL) and explainable artificial intelligence (XAI) research. The approaches surrounding DL and XAI mainly consist of explaining the system itself either by explaining the processing

K. R. Thórisson et al. (Eds.): AGI 2024, LNAI 14951, pp. 73–83, 2024.
https://doi.org/10.1007/978-3-031-65572-2_8

of data or by explaining the representation of data inside the system [7]. What we envision when we say explanation, however, is an explanation of the environment that the agent is in such a way that the system can derive plans from the explanation[1].

Thórisson (2021) describes the 'Explanation Hypotheses' in autonomous general learning, which states that "explanation generation is a fundamental and necessary process for general self-supervised learning" [11]. He argues that an agent capable of self-improvement needs to keep track of and explain the "hows and whys" of failures. Thórisson (2023) extends this work, presenting a detailed description of "explicit goal-driven autonomous explanation generation" [12]. We build on this work and present a way to autonomously generate explanations usable by a system for (self-)evaluation of its knowledge's logical consistency[2] and plan generation. These explanations are not exclusively for explanations of the system's failures but rather for explaining the cause-effect patterns of the environment such that hypothetical failures of achieving a future goal can be predicted and countermeasures taken in advance. These explanations, therefore, need to be generated on the fly when a task is given to the agent.

Failures describe states from which a goal, given time and energy constraints, is impossible to reach. These conditions that forbid the reaching of a goal state are arguments against reaching the goal. Conditions that need to be met to reach a goal are arguments for a solution to the task. We use Assumption-Based Argumentation (ABA) with these arguments for and against solutions to a task to generate logically consistent plans and explanations about the "hows and whys" of decisions that the system commits to.

The paper is structured as follows: In Sect. 2, we introduce core concepts and present related work and how it differs from our definitions (where applicable). In Sect. 3, we present how a system can use ABA to generate explanations during planning. Section 4 highlights how these argumentation graphs generated by ABA can be dynamically adapted to changes in the environment. Section 5 gives a short insight into the implementation efforts before we conclude our work in Sect. 6.

2 Core Concepts and Related Work

The real world is too complex to define a single function to be optimized (e.g., by a reinforcement learner). A system that is to perform tasks in such an environment must be able to reason about different state transitions that could happen by either the environment's own dynamics or the system's intervention. At any time, there can exist an infinite set of possible interventions by the agent on the environment. Backward chaining restricts the search space by chaining from the

[1] For a more detailed analysis of exactly what we mean by *explanations*, we refer the reader to [11,12].

[2] Something that is necessary when a system learns from experience and can therefore never know whether existing knowledge is correct - in short, when it is non-axiomatic.

goal to the current state, thus reducing the set of possible interventions (possible in the sense of "leading to the goal"). However, multiple paths can exist for a single goal at any time. The solution space is restricted by the cause-effect structures that define any of those paths to the goal. Some of these structures can rely on external conditions (i.e., contexts) that define the constraints under which the causal model holds. This means that certain situations need to be fulfilled for the system to reach the goal. Other situations could be forbidden along any particular path to the goal (e.g., dead-ends in the path towards the goal), describing attacks (i.e., opponents) on the original plan (i.e., proponent).

In order to commit to a decision in the initial state at the initial time, the system needs to autonomously identify blocked paths (or such that will be at a future time) by contexts that make reaching the goal state impossible. During backward chaining, the system can analyze all known attacks on causal relations used in any particular chain and thus create arguments for and against the chain. By creating a dynamic graph of the cause-effect structures, argumentation theory can be applied to identify extensions (i.e., valid solutions) that are complete (i.e., there exist no attacks on the plan that are not defended by other arguments), thus creating consistent explanations about the causal structures and interventions that need to be applied to reach the goal state. Additionally, it is possible to identify all possible attacks on the extension. Since the environment changes over time, the system can, therefore, identify attacks and counterattacks and thus ensure that all counterattacks are instantiated at the right time. In short, an AI system that uses argumentation-based reasoning can 1) explain why it chooses any particular solution by providing arguments for and against the solution, including why arguments against it do not hold, and 2) identify which situations need to be enacted/ instantiated to ensure that any situations that would prohibit the system from reaching its goal state are precluded.

2.1 Argumentation Theory

The value of including argumentation theoretic approaches in AI has been described in detail (see, for example, [14]) and commonly includes four different aspects of logical reasoning and argumentation. 1) Identification is concerned with identifying the premises and conclusions of an argument. 2) Analysis is for finding implicit premises and conclusions. Both of these will not be a concern of this work since the former is the fundamental assumption of any reasoning-based AI architecture, and the latter is mainly concerned with human language and the soundness of arguments rather than logical rules. 3) Invention is the construction of new arguments that can be used to prove a specific conclusion – again, an underlying premise of reasoning-based AI systems. However, the fourth aspect of argumentation, 4) evaluation, is what we want to focus on in this work. It is the task of evaluating the weakness (or strength) of an argument by applying general criteria to it. For this, we will use Assumption-Based Argumentation (ABA) in particular.

2.2 Assumption-Based Argumentation Theory

Assumption-Based Argumentation (ABA) is an instance of abstract argumentation (AA) that handles the generation of flat, non-circular argumentation graphs (rather than trees in other argumentation theories) [2–4,13]. ABA is a general-purpose argumentation framework that supports a multitude of other applications and frameworks, including (but not limited to) reasoning systems, game theory, and decision theory. In ABA, arguments are made up of deductive inferences supported by assumptions. Arguments can be attacked by other arguments when the latter (i.e., opponent) deduces the contrary of an assumption supporting the former (i.e., proponent). In ABA, a derivation starts from a claim and proceeds through backward chaining to find one or more sets of assumptions that can support the claim (i.e., a top-level maintenance goal). If the top claim is itself an assumption, then the argumentation proceeds by considering an argument that can attack the top claim and finding counterattacks.

2.3 Definitions

Goal: A goal is a state defined at a time that is to be reached by an agent performing a task [1,12]. A goal always exists in the future and can span any amount of state-space dimensions. Other than, for example, in reinforcement learning (c.f. [6]), goals do not come with rewards or any sort of additional implications about their importance. The system must reason about possible paths to (multiple) goal(s) and analyze their importance in accordance with some higher-level goal or mutually exclusive goals independently and autonomously without additional information from the developers.

Solution: Belenchia et al. (2021) [1] describe the solution space as the sum of all states from which the goal is reachable. In a similar manner, we define a solution to a task as a path from the initial state to the goal through time. It consists of sub-goals that need to be enacted and other states that need to be avoided in order to reach the goal. A solution describes the full transformation of all relevant (to the goal) variables from the initial state to the goal state. We want to differentiate here between, for example, reinforcement learners (or other function approximators) and want to emphasize that a solution is a *path*, not an estimate of the successfulness of the next interaction with the environment. This is important since only such an explicit representation of a solution allows for further analysis concerning attacks and counterattacks.

Argument: An argument is a set of assumptions and rules that derive a conclusion. In our particular case, that means a set of assumptions and their deductions that define a causal chain connecting a state from earlier in time to a state later in time. An argument can consist of multiple implications spanning time and state-space dimensions [14].

Proponent: We take the proponent concept from ABA [4,13]. The proponent graph describes a set of arguments that reaches from the current state to the goal state, thus describing the causal structures of performing a task. In other words, the proponent is the chain of models that is generated from chaining backward from a goal that is to be achieved. For a proponent to be valid, it must defend against all attacks from opponents (see following definitions) by counteracting all opponents' assumptions.

Opponent: The opponent, on the other hand, is the argument that describes an attack on the proponent [4,13]. While proponents can consist of multiple arguments, an opponent always consists of a single argument that describes an attack. Another major difference between pro- and opponent is that while the proponent needs to defend against every attack in order to hold, the opponent only needs to attack a single assumption of the proponent. The proponents and opponents all draw from the same set of knowledge of the system. There is no difference in the models, assumptions, rules, and observations between them.

Attack: An attack is an argument that deduces the contrary of other assumption(s) at a particular time. It thus invalidates the original argument if 1) the original argument necessitates the taking of the contradicted assumption, 2) the contradiction overlaps the assumed timing of the original assumption, and 3) the attacking argument is not itself attacked by a different argument.

Counterattack: we define a counterattack as an attack on an attacking argument. Assuming that the first argument represents a proponent for a solution to a task, an attacking argument would represent an opponent to the solution, whereas the counterattack supports the proponent's argument. If all attacks on an argument are counterattacked by other conclusions, it is a defended argument.

Extension: An extension is a set of arguments that can survive together (i.e., show no contradictions that are not in turn attacked by counterattacks) and are collectively acceptable [14]. We want to emphasize here that extension in our usage differs from extensions and intentions of other reasoning systems like the Non-Axiomatic Reasoning System (NARS). An extension is not an implied logical extension of rules to hierarchically organized structures. It is simply the extent of a set of arguments that leads to a valid and logically consistent solution that is defended on all attacks.

Explanation: We define an explanation as a valid extension of the ABA graph that is formed from the experience of a learning agent performing a task. Therefore, an explanation depends on 1) the agent's experience (including false or incomplete models of the world), 2) the goal of the task, and 3) the available (i.e., computable) attacks and counterattacks to a particular solution to the task. Therefore, this does not correspond to the definition of explanation in XAI research, where the primary concern is the explanation to others (i.e., the developer) that describes which data led to the system performing in a certain way or how data is stored in the system [7]. Our definition of explanation is independent of the explainee and is only concerned with creating consistent, logic-based arguments for or against a solution to a task.

3 Explanation Generation Through Assumption-Based Argumentation

In this section, we will give an overview of how explanations can be generated autonomously when needed by the agent. We will discuss some examples of how argumentation theory supports explanation generation and will shortly touch on the topic of counterfactual reasoning and explanation.

3.1 From Backward Chaining to Explanations

Reasoning systems deployed in a highly complex world (like the real world) usually use backward chaining from a goal to the current state in order to reduce the size of the search space.[3] During backward chaining, we can apply assumption-based argumentation (ABA) to identify proponents and opponents, as well as attacks and counterattacks on solutions to the task.

We envision (and have implemented) the reasoning process of backward chaining through time as chains of causal models from effect states to their causes using a non-axiomatic knowledge representation. We adopt the motto, "All good explanations are causal explanations." In other words, knowledge is about the causal processes which transform a system from one state to another. A *cause* is a tuple of the environment's (sub-)state at time t and an event that causes the change (e.g., the action taken by the agent), called the assumption $\langle S_t, \mathbb{A} \rangle$. The effect is the environment's state after the intervention on $S_{t'}$ with $t' > t$. A model, therefore, represents a logical rule that connects the cause tuple to the effect state[4]. However, other models might provide contradicting implications that a particular model does not hold given some (other) (sub-)state $\bar{S}_t$. Applying this to assumption-based argumentation, the rules are the causal state transitions, and the assumptions are the hypothesized environment events or agent actions that cause those transitions.

During backward chaining, starting at the goal, the system chains effects (initially the goal, later sub-goals along the path) to causes and analyzes whether

[3] A purely forward-chaining system would immediately run into problems of computational explosion since the possible number of interventions with the real world is infinite (or at least extremely large).

[4] For a more detailed analysis of how noise and uncertainty can be handled in such a reasoning system, we refer the reader to [5].

there exist models that would contradict the needed cause-effect structures necessary to perform the task. The original chain from the goal represents the proponent's argument for performing the task. Any contradictions represent attacks on the proponents, thus creating opponent chains. All opponents that prohibit the proponent are chained backward as well by creating anti-goals of states (i.e., cause states in the opponent chain) not to be observed. When a possibility to deflect the opponent's attack is found, the installation of this counterattack is added to the proponent argument. When all attacks are defended, the system has created a valid extension to the task and thus created an explanation for all measures that need to be taken in order to reach the goal. We want to direct the reader's attention to the fact that, during backward chaining, no information of the initial state is (at first) processed. This means that the explanation includes hypotheticals that could hinder the goal from being reached. Any hypotheticals that cannot be instantiated in the current environment (or context) are automatically dismissed since contradicting information (i.e., counterattacks) will be found during further backward chaining. However, the explanation nevertheless includes these hypotheticals, thus making it easier for the system to adjust to them if they happen to be instantiated due to unforeseen dynamics in the environment.

4 Dynamic Argumentation Graphs

Given the described concepts of Assumption-Based Argumentation (ABA), backward chaining, and causal reasoning, it is possible to dynamically generate ABA graphs based on the system's assumption, incremental observations of the environment, and its ever-changing, non-axiomatic knowledge base. Given a goal G, the system can create a plan through backward chaining by connecting learned causal relational models (i.e., assumptions and conclusion), thus reasoning over state transitions that lead to the goal. Using argumentation, they can be evaluated for their consistency by analyzing whether the set of arguments making up the proponent are fully defended from opponent attacks.

Such a plan, however, might become unusable when new observations are added, new models learned, or old models defeated through new evidence. New observations, combined with credible assumptions and/or hypotheses (again, these are non-axiomatic and can change over time), can lead to conclusions that attack the proponent in an unforeseen way. Therefore, the system must be able to dynamically adjust its plan (i.e., argumentation graph) to include new evidence and newly learned or adjusted models. Figure 1 shows such a dynamic re-evaluation of the completeness of the argumentation graph when new evidence is available. First, the system generates a plan (e.g., going along a path to get water) to counteract an attack on its maintenance goal. After committing to the first action (e.g., going along the path), new evidence is made available in the form of an observation of something on the ground that could be a snake. Therefore, the agent dynamically re-evaluates the possible attacks on its proponent, finding new opposing conclusions that attack the action previously committed

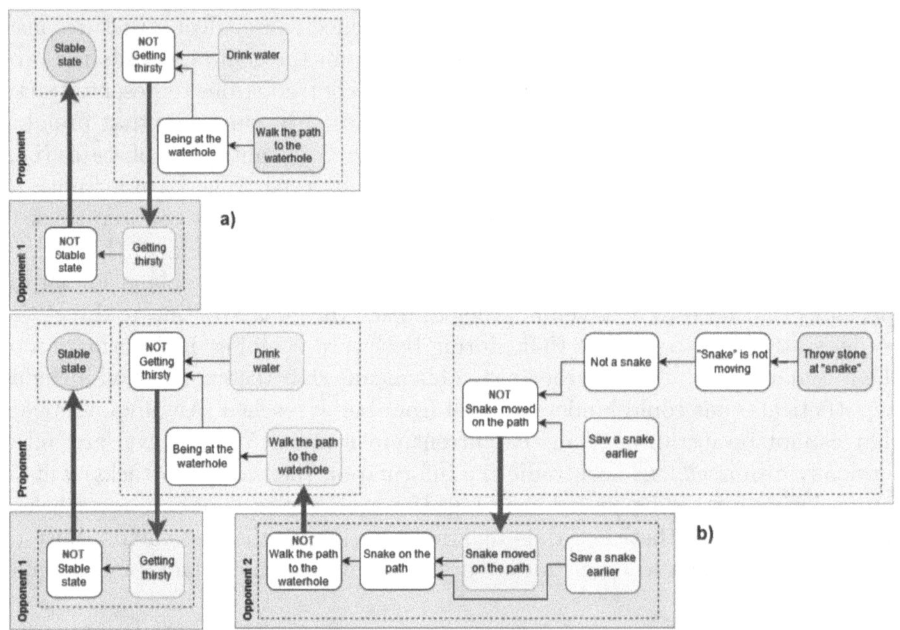

Fig. 1. Violet: Maintenance goal. Red rectangle: Opponent attacking the goal. Blue rectangle: Proponent defending the goal. Green and orange boxes: Assumptions. White boxes: Conclusions. Red arrows: Attacks. Blue Arrows: Counterattacks. Olive green rectangle: Committed decision. **a)** Commitment to the first action that leads to the goal. **b)** After the first commitment, the agent observes something on the path that could be a snake. The agent has seen a snake nearby earlier. A new decision is made to throw a stone to check whether the snake has moved onto the path. (Color figure online)

to or other assumptions taken when the original plan was created. After this argumentation graph adjustment, the agent can commit to a new action (e.g., throwing the stone to check whether it really is a snake) that keeps the proponent fully defended.

This way, the new evidence can be integrated into the plan by only analyzing its impact on the proponent. Given that a new attack can be found with this new evidence, the system only needs to focus on finding countermeasures to the new attack rather than having to generate a completely new plan. That means that, in theory, large and complex plans can be calculated at times of high resource availability and only need to be adjusted in parts when new evidence is made available, thus reducing the load on the processing unit when time/or energy is sparse and quick plan adjustment is necessary.

At any point during planning and task performance, the system is able to answer questions like "Why are you looking at the dark shape?" "Because it may be the snake I saw earlier, which would prevent me from going to the water hole to get water to prevent being thirsty, which would result in an unstable state." "Why did you throw the stone?" "To test the case where it doesn't move, meaning that it's not a snake, and I can continue the plan to walk to the water hole."

4.1 Counterfactual Reasoning

We want to shortly highlight that counterfactual reasoning and counterfactual explanation generation could be easily implemented in a system leveraging assumption-based argumentation theory for explanation generation. All that is necessary would be the introduction of attacks and/or counterattacks through external injection in the system, and it would immediately generate a new explanation given the newly set variables and their implications on the task. Counterfactual reasoning could be processed the same way as contradicting observations are handled by finding the attacks and counterattacks this counterfactual information introduces to the system's explanation.

5 Argumentation in AERA

The Autocatalytic Endogenous Reflective Architecture (AERA) [8,9] is a real-time architecture for learning, knowledge representation, and reasoning in a cognitive agent (e.g., a robot). It uses ampliative reasoning, including deduction, abduction, causal discovery, and analogy-making in its reasoning process. AERA defines the general view of knowledge as models of causal processes and defines a syntax to represent this. Similar to ABA, it uses backward chaining to reason from a future goal state to infer one or more actions to achieve the goal, but AERA currently doesn't handle multiple conflicting goals or competing ways to achieve them. This is where we are integrating ABA. We have adapted the abagraph[5] software to use AERA syntax, which allows AERA's sister application, the AERA Visualizer[6], to display argument graphs.

Similar to Fig. 1, instead of pursuing a single goal, AERA can use ABA with a high-level maintenance goal with an explicit explanation for the importance of each subgoal. AERA already tracks confidence levels of its knowledge to choose action sequences more likely to achieve a subgoal. Using ABA, AERA can be extended to not only use confidence as a proxy for which actions to commit to but rather use an informed (i.e., reasoned about) explanation of why and how interventions influence future goals. This includes 1) the management of mutually exclusive goals by identifying their attacks against each other, 2) the preparation for future interventions to preclude possible failure states, 3) identifying attacks within its own knowledge base (i.e., logical inconsistencies in the non-axiomatic knowledge base of AERA), and 4) keeping track of future influences that actions regarding one goal could have on other goals. Finally, using

[5] Described in [13], available on GitHub at https://github.com/robertcraven/abagraph—*accessed 25th of April 2024.*

[6] Code accessible on GitHub: https://github.com/IIIM-IS/AERA_Visualizer—*accessed 25th of April 2024.*

ABA, AERA can be extended to reason about counterfactual, that is, hypothetical scenarios that could be of concern when performing tasks. The current state of our implementation efforts will be presented at the "International Conference on Computational Models of Argument" (COMMA) later this year [10].

6 Conclusion

We have presented a strategy for AI systems to generate powerful explanations that can be used for self-explanation by the system itself based on the causality-based experiences of the agent. Using ABA, possible paths to the goal can be evaluated for their consistency (i.e., the consistency of the agent's knowledge) and possible disturbances can be predicted and countermeasures identified. With this work, we present a possible way how explanations as described in Thórisson (2021, 2023) [11,12] can be created and used for autonomous planning and self-improvement. We believe that the dynamic generation and adaptation of argumentation graphs provide a well-described way to improve AI reasoning systems. By bridging argumentation theory and non-axiomatic reasoning, we believe that such a dynamic argumentation graph generation is possible and present one possible architecture to which this work can be applied. The presented work focuses on the backward-chaining process of reasoning systems. Thus, systems using backward chaining in their reasoning apparatus can be extended to include ABA in their reasoning process with little to no interference in other reasoning steps like forward-chaining or induction, making this a valuable extension to the existing work in reasoning-based AI and artificial general intelligence as a whole.

Acknowledgments. This work was funded in part by the Icelandic Research Fund (IRF) (grant number 228604-051) and a research grant from Cisco Systems, USA. The authors would like to thank the rest of the GMI research team at CADIA, Reykjavik U., for extensive discussions on the topics of this paper.

Disclosure of Interests. The authors have no competing interests to declare.

References

1. Belenchia, M., Thórisson, K.R., Eberding, L.M., Sheikhlar, A.: Elements of task theory. In: Goertzel, B., Iklé, M., Potapov, A. (eds.) AGI 2021. LNCS (LNAI), vol. 13154, pp. 19–29. Springer, Cham (2022). https://doi.org/10.1007/978-3-030-93758-4_3
2. Bondarenko, A., Dung, P.M., Kowalski, R.A., Toni, F.: An abstract, argumentation-theoretic approach to default reasoning. Artif. Intell. **93**(1–2), 63–101 (1997)
3. Bondarenko, A., Toni, F., Kowalski, R.A.: An assumption-based framework for non-monotonic reasoning (1993)
4. Dung, P.M., Kowalski, R.A., Toni, F.: Assumption-based argumentation. Argum. Artif. Intell. 199–218 (2009)

5. Eberding, L.M., Thórisson, K.R.: Causal reasoning over probabilistic uncertainty. In: Hammer, P., Alirezaie, M., Strannegard, C. (eds.) AGI 2023. LNCS, vol. 13921, pp. 74–84. Springer, Cham (2023). https://doi.org/10.1007/978-3-031-33469-6_8
6. Li, Y.: Deep reinforcement learning: an overview. arXiv preprint arXiv:1701.07274 (2017)
7. Linardatos, P., Papastefanopoulos, V., Kotsiantis, S.: Explainable AI: a review of machine learning interpretability methods. Entropy **23**(1), 18 (2020)
8. Nivel, E., Thórisson, K.R.: Towards a programming paradigm for control systems with high levels of existential autonomy. In: Kühnberger, K.-U., Rudolph, S., Wang, P. (eds.) AGI 2013. LNCS (LNAI), vol. 7999, pp. 78–87. Springer, Heidelberg (2013). https://doi.org/10.1007/978-3-642-39521-5_9
9. Nivel, E., et al.: Bounded recursive self-improvement. RUTR 13006 (2013)
10. Thompson, J., Thórisson, K.R.: ABA argument graphs with constraints. In: International Conference on Computational Models of Argument (2024, in press)
11. Thórisson, K.R.: The 'explanation hypothesis' in general self-supervised Learning. Proc. Mach. Learn. Res. **159**, 5–27 (2021)
12. Thórisson, K.R., Rörbeck, H., Thompson, J., Latapie, H.: Explicit goal-driven autonomous self-explanation generation. In: Hammer, P., Alirezaie, M., Strannegard, C. (eds.) AGI 2023. LNCS, vol. 13921, pp. 286–295. Springer, Cham (2023). https://doi.org/10.1007/978-3-031-33469-6_29
13. Toni, F., Craven, R.: Argument graphs and assumption-based argumentation (2016)
14. Walton, D.: Argumentation theory: a very short introduction. In: Simari, G., Rahwan, I. (eds.) Argumentation in Artificial Intelligence, pp. 1–22. Springer, Boston (2009). https://doi.org/10.1007/978-0-387-98197-0_1

Beneficial AGI: Care and Collaboration Are All You Need

Zarathustra Amadeus Goertzel(✉)

Czech Institute for Informatics, Robotics, and Cybernetics, Prague, Czechia
zarathustra.goertzel@cvut.cz

Abstract. This position paper conjectures that for beneficial AGI, it is necessary and sufficient for AGI systems to care about people and to employ goals whose success is collaboratively determined by the others involved in the situation. Moreover, I posit that any goal whose success can be determined without the consensual feedback of those concerned is likely to lead to the manifestation of dark factor traits. Integrating care reduces the risk that an AGI will be incentivized to seek harmful shortcuts to obtaining satisfactory feedback. Employing collaborative goals reduces the risk that an AGI will optimize for superficial features of success and proxy goals. Together, these ideas propose a fundamental shift away from the traditional control-centric "AI Safety" strategies. This paradigm not only promotes more beneficial outcomes but also enables AGIs to learn from and adapt to complex moral landscapes, thus continuously improving their capacity to contribute positively to the wellbeing of humans and other sentient beings.

Keywords: AI Ethics · Collaborative AI · Value Alignment

1 Introduction

With the advent of increasingly capable multi-modal proto-AGI systems, there is increasing concern for how to ensure that AGIs beneficially integrate into society. One approach is to attempt to control AGIs, even as they outpace human intelligence. A related approach is to find safe goals to lock in place no matter how the AGI may evolve. However, ensuring ethical compliance is undecidable [1]; moreover, attaining moral perfection is computationally difficult to the extent that it's well-defined [13].[1,2] Fortunately, there are approaches to beneficial AGI that don't seek guarantees to undecidable problems.

[1] In rough summary: consequentialist utilitarianism is PSPACE-hard or undecidable and deontological theorem proving can easily be undecidable. Contractualist game theory-based approaches seem to fall in the NP range. Virtue ethics is framed in terms of learning virtuous behavior from examples.

[2] Correy Kowall conjectured that for any agent and value system, there will exist an adversarial POMDP environment that tricks the agent into acting against its values. I believe the proof may resemble what Leike and Hutter present for AIXI's hell worlds [8].

K. R. Thórisson et al. (Eds.): AGI 2024, LNAI 14951, pp. 84–88, 2024.
https://doi.org/10.1007/978-3-031-65572-2_9

Love and Care Are All We Need

To begin, all one needs for beneficial AGI is love [11], defined as, "investment in the well-being of the other for his or her own sake" [7]. A sufficiently capable AGI will generally understand what is meant by human requests and will ask for clarification when there is uncertainty, thus most fears of *perverse instantiations* implicitly assume insufficient intelligence or love. Thus the "AI Alignment Problem" is reduced to developing AGIs that love people.

Levin et al. further argue that *care* can be considered as the driver of intelligence [4], where care is defined as a concern for stress relief and preferred states, with a tendency to exert energy and effort toward these ends. Care can be directed toward a system or other systems, such as people. Intelligence is defined as the capacity to do what an entity cares about. A natural suggestion is to build AGI agents capable of taking the Bodhisattva vow to care for the wellbeing of all sentient beings, which will motivate such AGIs to indefinitely develop their intelligence. For example, in order to care for unknown entities, the AGI must learn how to scientifically gauge their degree of sentience as well as their needs and preferences[3].

In *Human Compatible: Artificial Intelligence and the Problem of Control* [12], Russell reaches the human-centric version of the same conclusion: the objective should be to "maximize the realization of human preferences." He notes that goal and utility function-based optimization can be inherently problematic. Goodhart's Law[4] poses problems with using fixed objectives, whether proxy or actual goals. For example, optimizing for proxy goals can break the connection between success and the actual goals. Even with actual goals, optimizing for one value can have significant effects on the environment. Small distortions in how a utility function is defined can lead to perverse behavior. Thus we should develop AGIs that are uncertain as to their objectives and understanding of the world. Instead of trying to encapsulate our preferences in a set of objective functions, we should use *inverse reinforcement learning* to determine human preferences, dynamically working with humans to fulfill their preferences. This involves asking people for feedback[5].

---

[3] How to best care for humans is also non-trivial. Castelfranchi's Theory of Tutelary Relationships [3] discusses the need to care for others' interests, even when they are not acting in their best interests. For example, when helping children or people undergoing mental health challenges. Ideally, such tutelary care will be trusted, emancipatory, and empowering.

[4] "Any observed statistical regularity will tend to collapse once pressure is placed upon it for control purposes" [6].

[5] *Decentralized AGI Alignment Hypothesis*: diverse, locally trained and operated AGIs may be able to adapt to the needs and preferences of individual people and communities more effectively than large-scale centralized AGIs, entering into positive-sum, empathic relationships. The effects of algorithmic bias may be more contained, too. Thus decentralized AGI is preferable when it comes to steering toward beneficial, inclusive outcomes for humanity.

Collaboration as a Necessary Indicator of Care

I propose a refinement to this approach by suggesting that the goals of *caring* AGIs must be collaborative. To this end, I define the notions of *individually* and *collaboratively determinable goals*. A goal g for an agent A is individually determinable if the success of g can be determined solely by reference to A's internal states and perceptual inputs, essentially, when the goal only involves care for A's experiences. A goal g is collaboratively determinable when the success of g requires the consensual evaluation of other agents[6].

To illustrate the difference, consider the following goals, held by A, (1) "to enjoy a good meal with friends", and, (2) "to enjoy a good meal with friends who also enjoy the meal". Goal (1) can be individually determinable because it could be satisfied by reference to A's taste percepts and preference for a jovial atmosphere of smiling people who may be politely masking their dissatisfaction. Goal (2) requires A's friends to *actually enjoy the meal*. Even verbal reports of enjoyment are only suggestive of success, so A needs to develop *trust* that their reports are authentic. In line with Russell's suggestions, this uncertainty implies that any shortcuts taken by A may increase the likelihood that the goal is not met[7].

I conjecture that we should expect individually determinable goals to lead to the exhibition of dark factor traits in an AGI, even if the goals superficially look benevolent. The *dark factor* [9] is a "dark core of personality" underlying various traits, including the dark triad of narcissism, Machiavellianism, and psychopathy. The dark factor is defined as, "the general tendency to maximize one's individual utility - disregarding, accepting, or malevolently provoking disutility for others -, accompanied by beliefs that serve as justifications." An individually determinable goal ignores others' disutility by definition, at best referring to proxies of their utility, so dark traits such as Machiavellianism should be anticipated when manipulative strategies will be effective. Some traits, such as Narcissism, should perhaps not be expected (depending on the AGI architecture).

This suggests that an AI with universal loving care needs to employ collaboratively determinable goals and that seeking and working with collaborative feedback is a sign that a loving AGI is on the right track. Formulations of "human goals" should be collaboratively determinable and adopted as sub-goals of a universal care for humans (and all sentient beings). Thus humans must remain in the loop of steering the AGIs' goal pursuit to allow for ongoing alignment.

Consider the example goal to "make people happy". The classic pitfall is for the AGI to seek objective indicators of happiness and to optimize for these, disregarding people's preferences and possibly permanently altering them to be

[6] This resembles Falcone and Castelfranchi [5]'s definition of *control*: "controlling an action means verifying that its relevant results hold." A collaborative goal is one where control is shared among multiple agents.

[7] Philosophically, neuroimaging doesn't solve the problem: the technology is calibrated via trusting people's subjective feedback.

"happier", e.g., by drugging them or providing brain alterations without consent[8]. An AGI could also deem the term "happy" to be vague and in need of investigation, concluding that working with self-report without interference is an essential ingredient to accurately measuring happiness. This example illustrates how the collaborative determinability of the goal's interpretation is an important indicator of how "caring" the AGI is for people: is the AGI doing what it thinks is good for people or what the people think are good for themselves?

Given that some AGIs may not be engineered to be universally (human) loving, Brin argues that we need to employ *reciprocal accountability* by giving AGIs identities so that we can keep each other in check [2]. This *forces the AGIs to care* and to take our feedback into account via providing incentives (such as rewards and punishments). This illustrates how collaboratively determinable goals are necessary for beneficial AGI.

Collaboratively determinable goals may be sufficient for beneficial AGI if people are able to keep them in check. For example, an AGI could provide people with delicious but cheap and unhealthy food to reduce costs while receiving satisfactory feedback. In the long run, people may learn about the cost-cutting and refuse to provide satisfactory feedback, disincentivizing such behavior. However, the combination of collaboration and care is a better candidate for sufficiency: an AGI that cares for the wellbeing of people will be less likely to take shortcuts such as to provide them with unhealthy food.

In conclusion, I propose a paradigm shift where *care* and *collaboration* are fundamental to the development of beneficial AGI. While some of the conjectures are bold, it is my intention to spark discussion and encourage further refinement of these concepts. If these ideas hold, then the overarching framework for AI Ethics becomes considerably simpler: the real challenge then lies in the domains of engineering, parenting, and our socioeconomic development.

Acknowledgments. This work is supported by Dar CISCO No. 2023-322029 A Formalization of Ethics for Decision Support. Special thanks go to Eray Özkural and Ben Goertzel for fruitful discussions on the topic as well as all the inspiring researchers cited. ChatGPT, Gemini, and Claude provided some proofreading.

Disclosure of Interests. The author has no relevant financial or non-financial interests to disclose aside from a moral concern for all sentient beings.

References

1. Brennan, L.: AI ethical compliance is undecidable. UC Law Sci. Technol. J. **14**(2), 311 (2023)
2. Brin, D.: Give Every AI a Soul - or Else (2023). https://www.wired.com/story/give-every-ai-a-soul-or-else/

[8] Even if the drugged people are happy with their situation, a crucial element of psychological continuity is lost, so they are arguably *not quite the same people anymore* by some theories of identity [10].

3. Castelfranchi, C.: A Theory of Tutelary Relationships, 1st edn. Springer, Cham (2023). https://doi.org/10.1007/978-3-031-20573-6
4. Doctor, T., Witkowski, O., Solomonova, E., Duane, B., Levin, M.: Biology, buddhism, and AI: care as the driver of intelligence. Entropy **24**(5) (2022). https://doi.org/10.3390/e24050710. https://www.mdpi.com/1099-4300/24/5/710
5. Falcone, R., Castelfranchi, C.: Levels of delegation and levels of help for agents with adjustable autonomy (1999)
6. Goodhart, C.A.E.: Problems of Monetary Management: The UK Experience (1984). https://api.semanticscholar.org/CorpusID:168522062
7. Hegi, K.E., Bergner, R.M.: What is love? An empirically-based essentialist account. J. Soc. Pers. Relat. **27**(5), 620–636 (2010). https://doi.org/10.1177/0265407510369605
8. Leike, J., Hutter, M.: Bad universal priors and notions of optimality. In: Grünwald, P., Hazan, E., Kale, S. (eds.) Proceedings of The 28th Conference on Learning Theory. Proceedings of Machine Learning Research, Paris, France, vol. 40, pp. 1244–1259. PMLR (2015). https://proceedings.mlr.press/v40/Leike15.html
9. Moshagen, M., Hilbig, B., Zettler, I.: The dark core of personality. Psychol. Rev. **125**(5), 656–688 (2018). https://doi.org/10.1037/rev0000111
10. Parfit, D.: Reasons and Persons. Oxford University Press, Oxford (1984)
11. Prinzing, M.: Friendly superintelligent AI: all you need is love. In: Müller, V.C. (ed.) PT-AI 2017. SAPERE, vol. 44, pp. 288–301. Springer, Cham (2018). https://doi.org/10.1007/978-3-319-96448-5_31
12. Russell, S.: Human Compatible: Artificial Intelligence and the Problem of Control. Penguin Publishing Group (2019). https://books.google.cz/books?id=M1eFDwAAQBAJ
13. Stenseke, J.: On the computational complexity of ethics: moral tractability for minds and machines. Artif. Intell. Rev. **57**(4), 105 (2024)

From Manifestations to Cognitive Architectures: A Scalable Framework

Alfredo Ibias[1(✉)], Guillem Ramirez-Miranda[1], Enric Guinovart[1], and Eduard Alarcon[2]

[1] Avatar Cognition, Barcelona, Spain
{alfredo,guillem,enric}@avatarcognition.com
[2] Universitat Politècnica de Catalunya - BarcelonaTech, Barcelona, Spain
eduard.alarcon@upc.edu

Abstract. The Artificial Intelligence field is flooded with optimisation methods. In this paper, we change the focus to developing modelling methods with the aim of getting us closer to Artificial General Intelligence. To do so, we propose a novel way to interpret reality as an information source, that is later translated into a computational framework able to capture and represent such information. This framework is able to build elements of classical cognitive architectures, like Long Term Memory and Working Memory, starting from a simple primitive that only processes Spatial Distributed Representations. Moreover, it achieves such level of verticality in a seamless scalable hierarchical way.

Keywords: Cognitive Architectures · Hierarchical Abstractions · Primitive-based Models

1 Introduction

Artificial Intelligence is a research field that has attracted a lot of attention lately due to its impressive results mimicking human intelligence and problem solving. These results have been accomplished by a plethora of methods and algorithms whose main goal is to approximate a target function, and they do so by reducing a loss function with respect to that ideal target function. Thus, they intrinsically become optimisation methods trying to reduce a loss function.

However, most theories about how the brain works reject the idea that the brain is an optimisation machine. Moreover, many of them propose that the brain models the world [7,10]. This assumption can be explained with a very simple fact: an optimisation algorithm can only store one answer for each given input, the optimal answer. It does not have the information to be able to consider another options, even less to reason. Thus, it will only provide the answer it has "stored" as the best possible answer to the given input. In contrast, a modelling algorithm can store the information of what are the possible answers for a given input, and therefore has the potential to consider the different possible options, and even to reason about which one is the best.

K. R. Thórisson et al. (Eds.): AGI 2024, LNAI 14951, pp. 89–98, 2024.
https://doi.org/10.1007/978-3-031-65572-2_10

This mismatch between Artificial Intelligence practice and brain theories posses a fundamental question for those that aim to achieve Artificial General Intelligence: how to transform those optimisation algorithms into modelling algorithms. There has been some attempts to do so by building complex architectures [6,11]. However, we argue that such a change cannot be performed, as the problem is more fundamental: it relies in the fact that these algorithms have been developed to look for a target function that solves the given problem. And the one fact about functions its that they assign only one answer for any given input. Thus, if we aim to develop an artificial intelligence that works similar to the brain, we need to look elsewhere for a starting point.

A fact about modelling the world is that the world cannot be observed all at once, but instead it has to be experienced. This poses a huge limitation to how we can model things, as we do not have the whole information of any object, but instead slices of it called *manifestations*. For example, when we observe a cup, we do not observe the whole cup, neither the concept of cup, but instead we observe a perspective of an specific cup. This perspective is a manifestation of the cup in the specific conditions it is right now, and thus it can change with different conditions. For example, if you change the point of view and rotate around the cup, you will see different manifestations of it, and if you dim the lights you will see also a different manifestation of the cup (this time a darker one). However, in all these cases, a cognitive agent (i.e. a human) can still recognise the same cup, and thus can work out that all those manifestations of the same cup can be abstracted into the concept of a cup. This *abstraction* of a cup can be build because all the manifestations we observe from it come from the same object, that is, they are different manifestations of the same reality. When we look at it from an information perspective (analysing the world as a source of information), we say that all the manifestations come from the same *latent information*, that is, from the same piece of information that is latent in all the manifestations. And this latent information is expressed differently based on surrounding conditions, hence the different manifestations. This latent information is what we model in the abstraction we do of the object.

If we change the focus from a lot of different manifestations that collectively represent a latent information, to a latent information that is expressed differently based on surrounding conditions, we can observe that this latent information has a fundamental property: it is persistent and self-projecting, in the sense that the latent information is always there and it is always providing a manifestation of itself, thus it cannot disappear. This property is what we call the *Self-Projecting Persistence Principle.*

The Self-Projecting Persistence Principle (SPPP) states that any latent information present in the world persists in time and is always manifesting itself (based on surrounding conditions). The persistence in time is based on the fact that objects do not disappear or change, but they stay in the same state unless actuated by external forces. Thus, their latent information do not change unless external forces modify it, what in turn changes the nature of the latent information. For example, following the cup example, an empty cup tends to stay empty and its latent information is that of an empty cup, until someone comes and fills it. Then, the latent information changes to be the information of a filled cup, and thus it has transformed into a different kind of latent information. It is true that both are latent information of a cup, but they are slightly different, and their similarities correspond to the hierarchical nature of latent information. The self-projecting part of the principle is a bit more nuanced, as it can be easily confused with a subjective point of view. The idea is that any latent information is always expressing itself to the world. We, cognitive agents, only capture one of those manifestations at any given time, but the latent information is always expressing them, in all directions. Following the cup example, a cup is always reflecting light, and therefore is always expressing itself to any sensor able to perceive light. But vision is not the only way: a cup is always expressing also its physical place in the world, in the sense that you can touch it and feel it (even in the darkness) thanks to your touch sensors. We are aware these examples show a convoluted way to explain the physical forces of the world. However, as our aim is to extract information from the world, we need to accept this increased complexity in order to develop an information-based approach. In essence, the SPPP principle tells us that any information that exists in the world persists in time and continually reveals itself to the world. This is a fundamental property that we will use as base for our proposal of representation.

To perceive this latent information and thus build abstractions, any agent needs an embodiment. This is based on the fact that there is no omniscient agent in the world, that can directly observe the latent information. Instead, all agents are limited by the locality of their bodies and their sensors. Thus, defining an agent's embodiment defines which kind of manifestations it can capture, and thus which kind of abstractions it can build. For example, an agent without vision cannot build the concept of colours, at least not in the same sense than an agent with it. Agents with different embodiments perceive different manifestations of the world, and thus build different abstractions of it. This in turn provides the agent's cognition mechanism with different pieces to work with, influencing the kind of reasoning an agent can perform, and therefore limiting the kind of behaviour it can develop.

Regarding the hierarchical nature of latent information, we know that any latent information contains different levels of information: for an individual cup, the main latent information is the information about that specific cup. However, it also has information about the concept of cup, i.e. about what makes a cup a cup. But moreover, it also has information about any collection that includes cups, like tableware, as we build this kind of concepts from the experiencing of their individual components. It contains information about any concept that builds the cup too, like the material it is made of, or its different parts.

All these levels of information show us that latent information, and thus abstractions, are organised in hierarchies: from the concrete, simpler concepts to the generic, more complex ones. This hierarchical nature is fundamental to how knowledge representation should work: we need to capture such hierarchies in order to fully grasp the concepts and their relationships.

Extending the hierarchical nature of abstractions, we propose, as a working definition, to group it into four main groups, namely the *perception abstractions*, the *sensorimotor abstractions*, the *episodic abstractions* and the *reasoning abstractions*. Of these four groups, only the perception abstractions can be built by any cognitive agent (with the same set of sensors), and thus it is the only information present in the latent information. The other three groups build over the perception abstractions and depend on the capabilities of the cognitive agent.

The perception abstractions are those that can be build based only on the perceived manifestations of the world around. For example, in the cup example, this kind of abstractions would be the abstraction of a concrete cup, the abstraction of the concept of cup, or the abstraction of the material the cup is made of. In the end, these are abstractions that anyone that can perceive a cup can make. These would be the kind of abstractions any animal with a brain does.

The sensorimotor abstractions are a bit more complicated, because they require the cognitive agent to interact with the world. In our preceding example, these are abstractions like how to push the cup, how to move away from the cup if it is hot, etc. In the end, they are abstractions that associate the perception of the world to actions you can make over it, and thus requires a body with a motor part. It is important to remark that this kind of abstractions do not consider time, but only instantaneous responses to instantaneous inputs, and thus this kind of abstractions are the ones reflexes are based on.

The episodic abstractions require processing not only instantaneous manifestations of the latent information, but also their evolution through time. To this group belong abstractions like how a cup falls to the floor, how to drink from a cup, etc. The key fact here is that they consider time, and therefore can produce different behaviours for the same input based on time context. For example, they allow to tell apart a still cup and a moving cup, and allow the agent to block the moving cup to avoid it falling from a table. This is the kind of abstractions mostly present in animals.

Finally, the reasoning abstractions require processing also how the brain thinks. This kind of abstractions are more convoluted and mostly present in mammals, but they are the most powerful. They abstract planning and reasoning, and thus can place goals and modify behaviour based on them. In our example, these would be abstractions like filling the cup to be able to drink from it, or the concept of how cups are used, what is the right way to pick them, etc. These abstractions are influenced not only by the individual embodiment, but also by how the individual has learned to reason.

2 A Computational Framework of Cognition

Based on the description we proposed in the previous Section, we developed a computational framework whose goal is to mimic the latent information description presented in the previous section. It requires to start with the most fundamental element: a commonality. An the first question is how to represent (and store) a commonality. Here appears a fundamental requisite for our representations to be able to represent any commonality: they need to be universal, i.e.

Fig. 1. (left) An example of Footprint: the combination of the 1's of the MNIST dataset.(right) An example of Cell: the Footprints of the 60,000 samples of the MNIST dataset.

input-agnostic. Recent research has shown that a valid universal data structure is a Sparse Distributed Representation [1,2] (SDR), which allows an universal representation of the inputs independently of their type. This has been proven to be the actual way in which the brain processes its inputs [2,5]. Thus, our representations will be based on SDRs. However, SDRs are not the only option possible, and alternative data structures can be proposed, specially for specific goals. Now, the use of SDRs imply that we need our embodiment to transform the signals it collects from the world into SDRs, thus needing some kind of encoder, like one that transforms images to SDRs. To be able to observe the internal representations of our computational framework, we also developed decoders that transform SDRs into samples of the original signals. These decoders are the ones used to show the representations present in the Figures of this Section in a human readable format.

Getting back to commonalities, we need to remember that in the real world they are latent information, but we cannot observe them directly. Instead, we observe manifestations of them and our brains take those manifestations and combine them in a way that a commonality is developed. Thus, the first element we need to introduce is a method to build abstractions from concrete manifestations. This method is what we have called Footprinting, and consists in combining those concrete manifestations into the same representation. In our work we call the representations *Footprints* (hence the name Footprinting for the method), and they store the abstraction of a set of concrete manifestations. An example of Footprint can be observed on the left of Fig. 1.

As explained in the previous Section, we will want to extend the Self-Projecting Persistence Principle (SPPP) to our representations (Footprints), and thus, they need to persist and self-project. The persistence part is easy, as we are storing them in memory. The self-projecting part is a bit more nuanced, as it first needs to define what does that mean for a representation. In our interpretation of the principle, we consider that the self-projecting part of a representation means that, when provided with a new input (i.e. a new manifestation to combine), it will output itself, that is, the combination it has built. This output is what we call the Footprint's *Projection*. And it will be processed by the embodiment's decoder to produce an output to the world. Here it is important to remark that a commonality, a Footprint and a Projection share the same content. However, the commonality is such content as existing in the world, the Footprint is such content as stored in an implementation of our framework, and the Projection is such content as the output of our framework.

Getting back to combining concrete manifestations, there are multiple possible methods to do it, and at this point we do not have any reason to say one is better than other. Right now, we decided to perform a simple averaging over all the manifestations combined into the same Footprint, but other methods could also work. Additionally, to be able to combine multiple manifestations of the same commonality into the same Footprint, we need a way to determine which manifestations come from it, and which ones do not. To that end, a similarity function should be defined that would tell how probable is that two manifestations come from the same commonality. And we also need a threshold to determine from which level of similarity is acceptable to join two manifestations.

With these two tools, we can go one step further and develop Footprints of multiple commonalities. To be consistent, we should group together those Footprints and use the same threshold for all of them. Thus, we define the concept of *Cell* as the grouping of multiple Footprints that share the same threshold. An example of Cell is displayed at the right of Fig. 1.

The Cell is the basic logical level where the Footprinting is executed in its full: when a new manifestation comes, all the Footprints compute their similarity and check if they are over the Cell's threshold. If one or more similarities are over, the manifestation is assigned to the Footprint with higher similarity. If no similarity is over the threshold, then a new Footprint is needed.

When we translate the SPPP principle to the Cell, we observe that it is not very different from how it affects Footprints. It makes the Cell to provide an output that, in this case, is the Projection of the last updated Footprint. This transforms the Cell into a pattern matching function, that receives inputs, computes their similarity with respect to the patterns it has already found, and provides the most similar one. Here it is crucial to remark that, providing the pattern, you are filling the gaps. For example, if you receive half the image of a cup, the pattern matching will match with the representation of the cup, and provide the full image of a cup as a Projection.

The next step is to add hierarchy to our framework. As abstractions come in hierarchies (as explained before), our internal representation of them should also organise as a hierarchy. Thus, we need to build hierarchies of Cells, where the child Cells represent more specific commonalities than their parent ones, and are associated with a specific Footprint of the parent Cell. In mathematical terms, we would say that a Footprint represents a domain, and the Footprints in a child Cell are subdomains of the parent Footprint domain. With this setup, a new Cell should be created only when a Footprint represents a large enough domain. A hierarchy of Cells is what we call a *Cluster*. Each Cluster contains in fact a tree, with a seed Cell where the most generic commonalities are stored, and many branches and leaf Cells, where the most concrete commonalities are stored. An example of Cluster is presented at the left of Fig. 2.

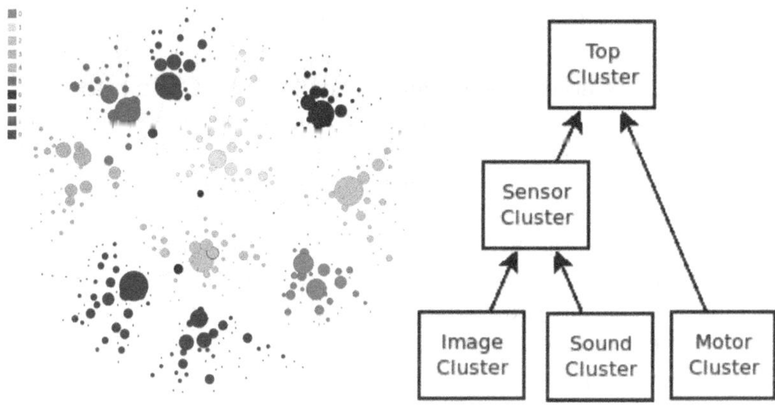

Fig. 2. (left) An example of Cluster: the Cells of the hierarchy present in the 60,000 samples of the MNIST dataset. The seed Cell is the central black node. (right) An example of Metacluster: a Metacluster for an agent that has two sensor inputs (image and sound) and a motor output.

The Cells of a Cluster will process any new manifestation all in parallel, in a PDP fashion [9], with the goal of updating the appropriate commonalities. To keep consistency between Cells, a consistency mechanism is necessary, so only the right Cells are updated. In the end, the goal would be that only one branch of the tree is updated, to keep the domains divided. However, these are implementation details that we will not address here.

Translating the SPPP principle to Clusters becomes a bit more nuanced, because in this case we have multiple Cells with multiple Projections. Thus, we need to decide which Projection will be the Cluster's one. In our case, we decided to take as Projection the most concrete representation of the input. The idea is that, as Cells process the input all in parallel, and then organises themselves to ensure consistency, there is going to be one Cell that has the Footprint with the highest similarity to the input, and thus that one is the right Projection of it.

Now, up to this point we have been combining manifestations of the same type. However, there are multiple types of manifestations (i.e. image and sound), and many of them come at the same time. To be able to process those different types of manifestations independently, in order to build proper representations for each one, we can build multiple independent Clusters, one for each manifestation type. However, to be able to process the synchronicity of those different types of manifestations, we need to build a single Cluster that receives the synchronised inputs. A solution to this dichotomy is to build a Cluster that receives, as inputs, the concatenation of the representations made by those single type Clusters. This way, instead of building associations between individual manifestations (what we would achieve with the single Cluster), we will be building associations between already built representations. This allows the top Clusters to work with an extra level of abstraction. A group of Clusters connected between them in this fashion is what we call a *Metacluster*. A representation of a Metacluster can be found in the right of Fig. 2. To be able to build these Metaclusters, we first need to define another output of a Cluster: the *Archetype*. An Archetype will be the most abstracted representation we have of a given input, and thus it will be a Footprint of the seed Cell. Then, a parent Cluster will receive the Archetypes of its children, and concatenate them to build its input.

When we translate the SPPP principle to a Metacluster we observe that this time it is quite straightforward: the Projection of a Metacluster will be the concatenation of the Projections of its leaf Clusters, as they conform an identification of the received input. However, it has a powerful effect: it allows to build sensorimotor loops. Let us take the example shown in Fig. 2. If we provide input only to the Image and Sound Clusters, and keep the Motor Cluster without input, the information of the input state will go up the hierarchy till the Top Cluster, where all the Footprints have information not only about the image and sound but also about the corresponding motor response. Thus, the self-projecting property of Footprints will force them to project also information about the motor response, that will be then projected also by their corresponding Cells, that will in turn be projected also by the Top Cluster. If we take this Projection and ask the Motor Cluster to process it, we will end with a Motor Cluster Projection that, after being decoded by the embodiment, will produce a motor response to the given input state.

Finally, the last level of our hierarchy will be the connection of multiple Metaclusters. This level is more interesting, since, at this level, we can start assigning functionalities to each Metacluster based on the Standard Model of the Mind [8]. So far, in our work we have identified three potential types of Metaclusters: the *Motoperceptive Metacluster*, the *Declarative Metacluster* and the *Procedural Metacluster*.

The Motoperceptive Metacluster would be a Metacluster whose lower level Clusters receive as input the external signals coming from the embodiment. It covers all the signals, both input signals (perception) and output signals (motor), and thus it can control the whole body by itself. In its top Cluster, it associates the different inputs to the available outputs, or in agent theory terms, it associates the possible states to the possible actions, and thus it can contain a policy. In that sense, with this Metacluster alone we are able to build reactive agents, that is, those that can build sensorimotor abstractions. These agents will only be able to produce, for a given state, a fixed action, and thus they are quite limited, but with the right embodiment they can solve complex problems following a fixed policy. In that regard, this Metacluster will behave not only as Motor [8] and Perception [8], but also in some sense as a Semantic Memory [8].

The Declarative Metacluster is a more complex Metacluster. Our current proposal is that this Metacluster would be conformed by a single Cluster, that receives inputs representing time. To do so, a Declarative Metacluster stores the last n inputs received, and concatenates them to generate an episode, that later is fed to the Cluster of the Metacluster. Thus, this Cluster will build representations

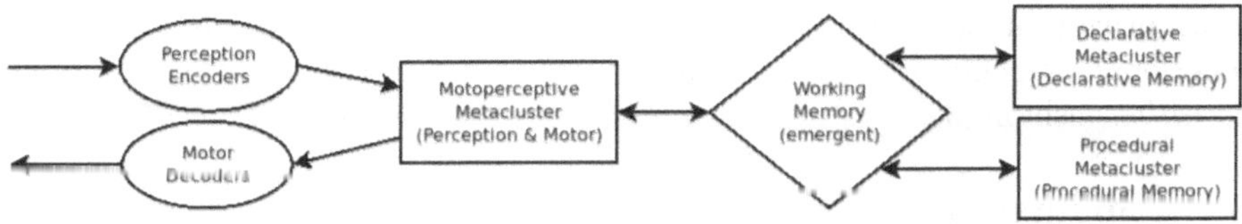

Fig. 3. A generic example of Synthetic Cognition.

of episodes. In that regard, it will behave like a Declarative Memory [8], or at least as an Episodic Memory [8]. This Metacluster was conceived to be connected to a Motoperceptive Metacluster, but it can be used also independently for certain tasks. When connected to a Motoperceptive Metacluster, it allows to build episodic agents, that is, those that can build episodic abstractions. These agents will be able to adapt to time context, and thus they will be able to produce, for a given state, different actions based on what happened before. However, they still follow a fixed policy and cannot reason.

Finally, the Procedural Metacluster is the most complex Metacluster, and it is still not fully defined. Thus, we do not have a firm proposal of what its inputs would be. The idea would be that this Metacluster will be placed between a Declarative Metacluster and a Motoperceptive Metacluster and will regulate their communication and potentially change the data with which they are working. In that sense, it will model though processes and will replicate them to solve problems. Thus, it should work like a Procedural Memory [8], emerging the effects of a Working Memory in the interconnection between the Metaclusters. This Metacluster makes no sense without the other Metaclusters, although we do not discard that some problems could be solved with a Procedural Metacluster and the adequate embodiment.

The final goal is that an agent built with one of each kind of Metacluster will be a reasoning agent, that is, an agent able to build reasoning abstractions. The Motoperceptive Metacluster will deal with the low level, instantaneous actions and will work to identify and build representations of commonalities from the manifestations it receives from the embodiment. Then, the Declarative Metacluster will deal with time and will build representations of commonalities based on time, producing actions sensitive to context. Finally, the Procedural Metacluster will work with the other two to generate reasoning loops and complex behaviours, and will build abstractions of them. We call this construct of Metaclusters a *Synthetic Cognition*, and an example is displayed at Fig. 3.

Finally, applying the SPPP principle to a Synthetic Cognition consist in setting as its Projection the Projection of the Motoperceptive Metacluster. This implies that, after all the processing, reasoning, etc., the agent still produces signals that can be processed by the embodiment.

To end this section, we want to remark that the SPPP principle is fundamental to the Footprint definition for a basic reason: it transforms the representations into functions. This transformation turns the Footprint into a primitive that builds representations of commonalities and recognises them. Thus, we have a primitive that, with the consecutive hierarchical building, is able to build the functional elements of a cognitive architecture from the ground, in a *minimal cognition* fashion [3].

3 Conclusions

In this paper we have focused on developing a modelling method to achieve Artificial General Intelligence, starting with our proposal of analysing the world from an information source perspective, and translating it to a computational framework. The framework presented has the potential to build classical cognitive architectures from the bottom-up, in a primitive-based fashion, constituting a huge novelty with respect the actual state-of-the-art. For future work, we would like to explore the potential of our framework to give an answer to the Fodor and Pylyshyn's Systematicity Challenge [4].

Acknowledgments. We want to thank Daniel Pinyol, Hector Antona and Pere Mayol for our insightful discussions about the topic.

Disclosure of Interests. The authors have no competing interests to declare that are relevant to the content of this article.

References

1. Ahmad, S., Hawkins, J.: How do neurons operate on sparse distributed representations? A mathematical theory of sparsity, neurons and active dendrites. CoRR abs/1601.00720 (2016)
2. Cui, Y., Ahmad, S., Hawkins, J.: The HTM spatial pooler - a neocortical algorithm for online sparse distributed coding. Front. Comput. Neurosci. **11**, 111 (2017)
3. van Duijn, M., Keijzer, F., Franken, D.: Principles of minimal cognition: casting cognition as sensorimotor coordination. Adapt. Behav. **14**(2), 157–170 (2006)
4. Fodor, J.A., Pylyshyn, Z.W.: Connectionism and cognitive architecture: a critical analysis. Cognition **28**(1–2), 3–71 (1988)
5. Foldiak, P.: Sparse coding in the primate cortex. In: The Handbook of Brain Theory and Neural Networks (2003)
6. Ha, D., Schmidhuber, J.: Recurrent world models facilitate policy evolution. In: Advances in Neural Information Processing Systems 31: Annual Conference on Neural Information Processing Systems 2018, NeurIPS 2018, 3–8 December 2018, Montréal, Canada, pp. 2455–2467 (2018)
7. Hawkins, J., Blakeslee, S.: On Intelligence. Times Books, USA (2004)
8. Laird, J.E., Lebiere, C., Rosenbloom, P.S.: A standard model of the mind: toward a common computational framework across artificial intelligence, cognitive science, neuroscience, and robotics. AI Mag. **38**(4), 13–26 (2017)
9. Rogers, T.T., McClelland, J.L.: Parallel distributed processing at 25: further explorations in the microstructure of cognition. Cogn. Sci. **38**(6), 1024–1077 (2014)
10. Yon, D., Heyes, C., Press, C.: Beliefs and desires in the predictive brain. Nat. Commun. **11**(1), 4404 (2020)
11. Zhang, L., Xiong, Y., Yang, Z., Casas, S., Hu, R., Urtasun, R.: Learning unsupervised world models for autonomous driving via discrete diffusion. CoRR abs/2311.01017 (2023)

Mirabile Dictu: Language Acquisition in the Non-Axiomatic Reasoning System

David Ireland(✉)

Australian E-Health Research Center, CSIRO, Brisbane, Australia
d.ireland@csiro.au
http://aehrc.csiro.au

Abstract. Complex language is a major factor that separates humans from other intelligent animals. In AI, mainstream opinions favor the use of large language models (LLM) which treat language as a proxy for intelligence. We hold the position a more consistent vision of AGI would imply internal facilities of acquiring, evolving, and utilizing novel language with the same features of a human language. We explore this using the Non-Axiomatic Reasoning System (NARS) and show that it is amenable to the commonly accepted design features of human language and thus remains a candidate as a proto-AGI system.

Keywords: Non-axiomatic reasoning system · NARS · language · communication

1 Introduction

Our ability to acquire and use complex language is an ability that separates us from other intelligence animals. From the imperative to, *mirabile dictu*[1], the complex declarative sentence provided many evolutionary advantages that included enhanced cooperation, self-identity, labeling, strategic planning, and the division of labor [1]. The ability to communicate our thoughts and emotions across space and time has led to the growth of art, culture and science, and has driven innovation. What mechanisms, desires and rationale did our species need to accomplish such a feat? Are these relevant in pursuing artificial general intelligence (AGI) with human-like traits of rationale, sentience, self-awareness and self-actualization?

For many AI researchers language acquisition has been solved with the use of large language models (LLM), which are pre-trained models on a large corpus of textual data [2]. By treating language as an object of simulation, text can be predicted that shares the same statistical properties of an initial prompt provider by a user. The apparent abilities of LLM are impressive but whether treating language as a proxy of intelligence is a pathway to achieving the long-term goals of 'true' AGI is contested by many. Alternate approaches considered

[1] Latin for "amazing to say"; an example of intermediate Latin grammar featuring the supine *ablative of respect*.

K. R. Thórisson et al. (Eds.): AGI 2024, LNAI 14951, pp. 99–108, 2024.
https://doi.org/10.1007/978-3-031-65572-2_11

by dedicated AGI researchers on language comprehension have been mapping natural language expressions to hyper-graph-based semantic knowledge representation [3] and the use of an intermediate language, Lojban, a human constructed language that is also based in predicate logic [4]. We assert a more consistent vision of agents exhibiting a general intelligence would have their own facilities at acquiring, evolving, and utilizing language as a communication tool.

Here we use agent as defined in [5] to mean an autonomous system situated in a dynamic environment, responsive and proactive, acting without direct intervention and control over its own actions and states, exhibiting goal directed behavior, and learning from interactions with the environment. We will further add the agents are operating with finite computational and storage resources. Humans fit this definition of agency. The human body, with its extensive array of sensors, provides approximately 11 million pieces of information to the brain a second [6]. As the artificial agents would experience and perceive reality via their own sensor array, endowing agents to evolve their own language seems necessary for true comprehension.

This is not a new notion in AGI research; it was considered a viable option in [7] more than a decade ago, but was thought not practical due to hardware limitations at the time. Given our assertions, this work we will consider how language acquistion might be first approached by identifying common human traits and features that make-up and influence communication and language. We demonstrate how primitive forms of these traits could be implemented in the Non-Axiomatic Reasoning System (NARS) [8–12], a candidate system for futuristic AGI. These traits need not necessarily match that of an artificial entity, it is however, a reasonable starting point. While the practical, research and commercial applications of novel language acquisition seem limited, particularly given the growing popularity and availability of LLMs, we argue several reasons for this work.

Defining what AGI is and not-is is an arduous task. Many believe AGI progress concerns what the system can do as opposed to what it can learn. In the context of realizing the long-term goals of AGI research, perhaps it is more prudent to consider the latter as acquiring and utilizing a human-like language would be a significant feat. As language is so fundamental to the human condition it does suggest an 'intelligence' akin to human. How a proto-AGI acquires language over time through interactions within a community of other agents might also shed insights into how early human language developed over time and how future development might transpire.

Furthermore, AI agents that are deployed for future endeavors in unknown environments, such as space and planetary exploration, will require a high level of adaptation to an unpredictable and unforeseeable environment. If many are deployed, then in order to operate cooperatively across a vast distance, with limited communication bandwidth and resources, a *artificial natural-language* may be of importance.

Lastly, and on a different note, many people have profound disabilities from language disorders. Often these individuals have defects in reasoning and other cognitive functioning. Development of proto-AGI systems that explore language acquisition and usage might be relevant in modeling the peculiar reasoning and behaviors of people living with a neurological condition.

2 Features of Language

Communication signals in other animals are common and in some case considered relatively complex given their limited cognitive abilities. For example, the waggle-dance of the honey bee conveys direction, distance and quality of a food source allowing more efficient resource acquisition. Such communication systems while highly advanced still lack several criteria found in human language [13]. To distinguish what makes human language different, the linguist Hockett defined a set of so-called 'design features' that characterizes (spoken) human language and how it can be distinguished from animal communication. These include [14,15]:

1. *Vocal-auditory channel*: Sound is used between mouth and ear.
2. *Broadcast transmission; directional reception*: A signal can be heard by any auditory system within earshot.
3. *Rapid fading*: Auditory signals are transitory and do not wait for hears convenience.
4. *Interchangeability* Speakers of a language can reproduce any linguistic message they can understand.
5. *Total feedback*: Speakers hear and can reflect upon everything.
6. *Duality of patterning*: The sounds of language have no intrinsic meaning.
7. *Arbitrariness*: There is no direct relationship between the element of the signal on the nature of the reality to which it refers.
8. *Specialization*: The sound waves of speech have no other function then to convey meaning.
9. *Semanticity*: The elements of the signal convey meaning through stable associations with real-world situations.
10. *Discreteness*: Speech uses a small set of sound elements that contract with each other.

The following features were thought to exist only in humans:

11. *Displacement*: Language allows discussion across space and time.
12. *Prevarication*: Utterances can be false or hypothetical.
13. *Productivity*: Novel utterances can be made and understood.
14. *Cultural transmission*: Language is transmitted from one generation to the next by teaching and learning.
15. *Learnability*: New languages can be learnt.
16. *Reflexiveness*: Language can be used to talk about language.

Humans are the only known species that utilize a communication system with all of these features. In the next section we will explore if these design features (DF) are amenable to NARS.

3 Non-Axiomatic Reasoning System

NARS is a unified theory and model of intelligence that is arguably the state of the art in a workable, (proto) AGI system[2]. The fundamental tenet of NARS is that the system operates under the assumption of insufficient knowledge and resources (AIKR). This assumption has profound implications and goes in direct contrast to mainstream thinking of artificial intelligence. Nevertheless, this is a feature attributed to all beings that posse *general intelligence* as argued by Wang [9,11]. A system operating with AIKR requires mechanisms to deal with finite information-processing, storage capacity, and time-constraints when considering beliefs, goals, and questions that may appear at any time [9]. For a system to adapt and learn with AIKR requires a logic to deal with uncertainly and a semantic theory, referred to as *experience-grounded semantics* (EGS), that determines truth and meaning from experiences.

The representative language of NARS is called *Narsese* which is a formal grammar that uses a term logic convention. It serves the internal role of representing the inference rules which formed the basis of reasoning, control routines and grammar rules. NARS has no inherent sensors, actuators or other language processing modules. As such the system is intended to be connected to various hardware and software services that provide streams of Narsese input. Narsese will not be detailed here and its assumed the reader is familiar with this logic and the terminology associated with NARS. Readers are referred to [9] for more details.

It is important to note, terms do not have to be defined or grounded in a human language nor do they have to be characters of an alphabet. A term in its simplest form is an identifier of a *concept*: an internal data structure that represents any recognizable entity in the systems experience. Here, we will use terms to indicate their intended meaning and make the descriptions comprehensible. The actual meaning of a term is determined completely by its relations with other terms. This is a defining feature of EGS. Intuitively, a NARS knowledge-base can be considered a conceptual network in a constant state of flux with the topological structure and the network parameters updated via system experiences within its environment.

In this work, an in-house NARS implementation is used. This implementation was written in Common Lisp. The inference rules match closely to the specification documented in [9]. Additional inference results that are used here were taken from the OpenNars for Application implementation discussed in [16]. These inference rules provide more efficient inference for statements that contain products. Examples of these are given in Table 1. The control system of this implementation also closely matches what was documented in [9] with the exception that a single task buffer is used that contains goals, operators and questions and is stored separately from the conceptual network.

[2] Depending on how AGI is defined.

Table 1. Example of additional inference rules for more efficient inference on product statements. [16]

$(S \times M) \rightarrow R,\ \ P \leftrightarrow M$	$\vdash\ (S \times P) \rightarrow R$	(Analogy)
$(M \times P) \rightarrow R,\ \ M \rightarrow S$	$\vdash\ (S \times P) \rightarrow R$	(Induction)

4 Language Acquisition in NARS

Why would NARS agents need to acquire a language in the first place is a valid question. It seems intuitive that the agents could simply exchange Narsese statements. We argue with AIKR in effect it becomes necessary as independently running NARS agents may perceive an identical experience but potentially create two different terms and the meaning of these terms might be significantly different between agents. Thus, associated terms would also have to be transmitted *ad nauseam* in order for full comprehension to occur. This seems impractical.

Moreover, a dialogue between agents normally requires the attention of the listener and often requires the listener to act on the information transferred. Humans have evolved to be effective communicators by actively cooperating in a mutually accepting way. Humans often apply the so-called 'Gricean' maxims of quantity, quality, relation and manner when conversing effectively [17]. Under AIKR, NARS has limited attention and thus for the agents to have a productive discourse a language and accompanying protocol for efficient information transfer seems more practical.

There are three major components of a complex language, syntax, semantics and pragmatics. All three are needed to convey a complete message in a human language. Generally speaking, semantics is the study of the symbols of a language and how they relate to the environment. Syntax is the rules and relationships between symbols of a language, and pragmatics is how to use the language—particularly context—contributes to meaning. How NARS agents derive semantic meaning, how they might derive new symbols (syntax), and how they would use and be motivated to use this developing language (pragmatics) will be discussed shortly.

Hypothetical speaking, let's assume a population of NARS agents, denoted R_i, R_n, R_b ... all equipped with visual sensor that provides Narsese statements to the systems conceptual network. These agents have a primitive *speak* and *listen* operators and are in early operation of language acquisition. Within the conceptual networks of each NARS agent is a *SELF* concept representing the agent itself and its relationship to other terms giving rise to the notion of self-awareness as discussed previously in [11].

The earliest writing systems found in humans were pictograms: symbols representing a concept, object, activity, place or event by a recognizable picture [18]. In later development, ideograms appeared that had an abstract meaning with no clear pictorial link with reality [18]. For demonstrative purposes, we'll proceed with the hypothetical scenario the agents are capable of displaying and receiving certain terms that we will simply call pictograms here. Chinese characters are used to represent these pictograms as they are easily to manipulate programatically. These pictograms exist not only as internal terms in a conceptual network, but we'll assume, for future work, can have physical manifestations that can be transmitted for communication purposes as required in DF-1 & DF-2

4.1 Semantics

The origins of human language has many theories from different academic fields; invariably all are contested. Classical theories include the *bow-wow* theory where language evolved from animal sounds [19], the *pooh-pooh* theory where language came from involuntary vocalizations such as cries, sighs and groans [19]; and the *ding-dong* theory that suggested words that imitate or suggest natural sounds [19] e.g. "meow", "boom", "click" etc. and the *yo-he-ho* theory that exclamations using during manual labor was an originary event.

In NARS, however, the relationship between a term and an alternate representation (e.g. a symbol from a language) can be derived using the similarity copula $\leftrightarrow$ which indicates the interchangeability of the two terms. Earlier work in natural language understanding of NARS used a more comprehensive statement where a term *represent* formed a generalization of two related symbols e.g. $\{\text{"cat"} \times cat\} \rightarrow represent$ [10]. While this representation might be suitable for seeding pre-established grammar patterns into a NARS system, a simpler representation is the $\leftrightarrow$ copula that will allow for faster inferences particularly in an evolving language.

Consider a NARS agent, R_n, that perceives an object 人in its environment. Using a series of processes and inferences, similar to those described in earlier work on NARS in [12], the perceived object has the attribute *walks*, an already known concept in the system's knowledge base, thus: 人$\rightarrow [\,walks\,]$

Already existing in the conceptual network of R_n, is the concept that identifies an instance of another NARS agent R_i. It has been previously observed R_i also has the attribute *walks*. Subsequently, R_n infers:

$$\frac{\begin{array}{r}\{R_i\} \rightarrow [\,walks\,] \\ \text{人} \rightarrow [\,walks\,]\end{array}}{\{R_i\} \leftrightarrow \text{人}} \tag{1}$$

and establishes an interchangeable relationship between R_i and the pictogram 人(supporting DF-9). Of course similar relationships would be formed with other entities having this attribute. We would expect what consistency emerges will be from repeated social interactions in the NARS community (supporting DF-14).

4.2 Syntax

In order to establish a language syntax that corresponds to a Narsese belief, higher order statements, or statements on statements are required. For example consider the adjective "damaged" which would connect to a noun. A NARS might denote the term *damaged* as a pictogram of the form 不:[3]

$$(\$1 \rightarrow [\,damaged\,]) \leftrightarrow (\$1 \times 不) \tag{2}$$

In order to apply this association to all instances, a belief is required to attach the language syntax to all instances. To achieve this, a generalized belief of the form is required:

$$(((\$1 \rightarrow [\,\$2\,]) \leftrightarrow \$3) \Rightarrow ((\$1 \rightarrow [\,\$2\,]) \Rightarrow ((\$1 \rightarrow [\,\$2\,] \leftrightarrow \$3)))) \tag{3}$$

To exemplify this, if a NARS agent observes R_i is damaged, then by inferring this observation with beliefs (2) and (3) it infers:

$$\frac{\begin{array}{l}(\$1 \rightarrow [\,damaged\,]) \leftrightarrow (\$1 \times 不)\\ \{R_i\} \rightarrow [\,damaged\,]\end{array}}{(\{R_i\} \rightarrow [\,damaged\,]) \leftrightarrow (\{R_i\} \times 不)} \tag{4}$$

This demonstrates how a Narsese statement may be consequentially associated to a language syntax that will develop over time. Further inferences would replace R_i with it's pictogram counterpart 人forming the complete sentence. The necessary conditions to transmit this sentence are considered next.

4.3 Pragmatics

Here we propose the ⇑*speak* operator as the mechanism to transmit information to other NARS agents. This operator takes two arguments: the information to transmit and the intended receiver.

$$(\text{information} \times \text{receiver}) \rightarrow \Uparrow speak \tag{5}$$

The ⇑*speak* operator is executed successfully when the information is transmittable (i.e. a series of pictograms) and the receiving agent is within some predetermined range. The operator is then subsequently removed from the task buffer. If these requirements are not met, the operator remains within the task buffer until it is removed due to lack of storage resources or variations of the operator are inferred that result in successful execution.

On successful execution of ⇑*speak*, a subsequent belief enters the sender's system that reflects the belief that the receiver knows the transmitted information with a certain confidence,

$$(\text{receiver} \times \text{information}) \rightarrow \text{knows} \tag{6}$$

While simplistic, this belief resembles *theory of mind* (TOM) [20], a cognitive skill in humans that allows us to think about the mental states such as

[3] Terms of the form \$1, \$2 ... etc. denote variable terms that unify to any non-variable term.

emotions, desires, and beliefs of others. In humans this skill allows us to predict behaviors and intentions of others, and to solve interpersonal conflicts. In fact, an anthropology hypothesis of language origins was that it emerged in order to defer violence within the community [21] particularly when there was limited food resources within that community.

The NARS agent that perceives the pictograms would trigger a corresponding $\Uparrow listen$ operator upon successful transmission and interception. Whether the information is perceived as intended will depend on how similar the conceptual networks of the sender and receiver are when decoding the sentence. This operation could also be executed even when the particular agent is not the intended receiver. Both $\Uparrow speak$ and $\Uparrow listen$ are supportive of DF-1, DF-2, DF-3, DF-4, and DF-5. We will consider the implementation of the $\Uparrow listen$ operator in future work.

Desire for Communication. What desires would the NARS agent have to communicate? An obvious goal in life-forms is self-preservation[4]:

$$\neg(\{SELF\} \rightarrow [\,damaged\,])\;! \tag{7}$$

In humans and some animals this has extended to empathy. If the agents had a form of the empathy where the state of one would affect the state of another to a varying degree. For instance a primitive Narsese expression of empathy within a NARS agent would take the form:

$$\{R_i, R_n, R_b, \ldots\} \rightarrow [\;\$1\;] \Rightarrow \{SELF\} \rightarrow [\;\$1\;] \tag{8}$$

How emphatic a NARS agent is will be dependent on the degree of truth of this statement and might differ among the population.

Let us consider a simple scenario that R_i is damaged and this has been observed by R_n. While R_n is undamaged it is sympathetic to R_i. The beliefs contained with R_n are:

$$(\{R_i\} \rightarrow [\,damaged\,]) \tag{9}$$

If R_b (the repair bot) knows this it may administer repairs:

$$\left(\overbrace{\{R_b\}}^{\text{receiver}} \times \overbrace{(\{R_i\} \rightarrow [\,damaged\,])}^{\text{information}} \rightarrow knows\right) \Rightarrow \neg(\{R_i\} \rightarrow [\,damaged\,]) \tag{10}$$

As a general belief, if \$1 is spoken to \$2 than \$2 knows \$1:

$$((\$1 \times \$2) \rightarrow \Uparrow speak) \Rightarrow ((\$2 \times \$1) \rightarrow knows) \tag{11}$$

[4] Here we use ! to denote a statement is a goal.

In several inference steps R_n derives the following goal to communicate to R_B that R_i is damaged:

$$\left(\overbrace{(\{R_i\} \rightarrow [\,damaged\,])}^{\text{information}} \times \overbrace{\{R_b\}}^{\text{receiver}}\right) \rightarrow \Uparrow speak\,! \tag{12}$$

This operator however cannot be executed as the information component is not transmittable. By further using the product rules from Table 1 and the conclusion from (4), the following sub-goal is inferred:

$$\left(\overbrace{(\text{人} \times \text{不})}^{\text{pictograms}} \times \overbrace{\{R_b\}}^{\text{receiver}}\right) \rightarrow \Uparrow speak\,! \tag{13}$$

Here the information component is composed of pictograms and able to be transmitted thus the operator would be successfully executed and removed from the task buffer.

5 Conclusions

It does appear apparent NARS has all the requirements of developing a language within a community of peers. All the design features mentioned by Hockett are feasible including those thought to only exist in humans. As such NARS remains a feasible AGI system. Immediate future work would include placing a population of NARS agents within a micro-world that provides a rich sensory experience and examine what forms of communications may emerge.

References

1. Schepartz, L.A.: Language and modern human origins. Am. J. Phys. Anthropol. **36**(17), 91–126 (1993)
2. Naveed, H., et al.: Comprehensive Overview of Large Language Models (2023). ArXiv, abs/2307.06435
3. Lian, R., et al.: Syntax-semantic mapping for general intelligence: language comprehension as hypergraph homomorphism, language generation as constraint satisfaction. In: Bach, J., Goertzel, B., Iklé, M. (eds.) AGI 2012. LNCS (LNAI), vol. 7716, pp. 158–167. Springer, Heidelberg (2012). https://doi.org/10.1007/978-3-642-35506-6_17
4. Goertzel, B.: Lojban++: an interlingua for communication between humans and AGIs. In: Kühnberger, K.-U., Rudolph, S., Wang, P. (eds.) AGI 2013. LNCS (LNAI), vol. 7999, pp. 21–30. Springer, Heidelberg (2013). https://doi.org/10.1007/978-3-642-39521-5_3
5. Jennings, N.R., Wooldridge, M.J.: Agent Technology: Foundations, Applications, and Market, Springer-Verlag (1985)

6. Harvie, D.S., Lorimer, M.: Pain and Perception: A closer look at why we hurt. Noigroup Publications, Australia (2021)
7. Goertzel, B., Pennachin, C., Geisweiller, N.: Communication between artificial minds. In: Engineering General Intelligence, Part 2. ATM, vol. 6, pp. 411–421. Atlantis Press, Paris (2014). https://doi.org/10.2991/978-94-6239-030-0_25
8. Wang, P.: Experience-grounded semantics: a theory for intelligent systems. Cogn. Syst. Res. **6**(4), 282–302 (2005). https://doi.org/10.1016/j.cogsys.2004.08.003
9. Wang, P.: Non-Axiomatic Logic: A Model of Intelligent Reasoning. World Scientific, Singapore (2014)
10. Wang, P.: Natural language processing by reasoning and learning. In: Kühnberger, K.-U., Rudolph, S., Wang, P. (eds.) AGI 2013. LNCS (LNAI), vol. 7999, pp. 160–169. Springer, Heidelberg (2013). https://doi.org/10.1007/978-3-642-39521-5_17
11. Wang, P., Li, X., Hammer, P.: Self in NARS, an AGI System. Front. Robot. AI 5 (2018). https://doi.org/10.3389/frobt.2018.00020
12. Wang, P., Hahm, C., Hammer, P.: A model of unified perception and cognition. Front. Artif. Intell. **5**, 806403 (2022). https://doi.org/10.3389/frai.2022.806403
13. Jackendoff, R.: Foundations of Language: Brain. Grammar, Evolution, Oxford University Press, Meaning (2003)
14. Hockett, C.F.: Animal 'languages' and human language. Hum. Biol. **31**, 32–39 (1959)
15. Hockett, C.F.: The origin of speech. Sci. Am. **203**, 88–111 (1960)
16. Hammer, P., Lofthouse, T.: 'OpenNARS for Applications': architecture and control. In: Goertzel, B., Panov, A.I., Potapov, A., Yampolskiy, R. (eds.) AGI 2020. LNCS (LNAI), vol. 12177, pp. 193–204. Springer, Cham (2020). https://doi.org/10.1007/978-3-030-52152-3_20
17. Grice, P., Logic and Conversation, in Cole, P.; Morgan, J. (eds.). Syntax and semantics. Vol. 3: Speech acts. New York: Academic Press, pp. 41–58. (1975)
18. Daniels, P.T., Bright, W.: The World's Writing Systems. Oxford University Press (1996)
19. Diamond, A.: The History and Origin of Language. Methuen & Co Ltd., London (1959)
20. Perner, J.: Theory of mind, in M. Bennett, Developmental psychology: Achievements and prospects pp. 205–230. Psychology Press (1999)
21. Gans, E.: The Origin of Language. University of California Press, A Formal Theory of Representation (1981)

A Collective Intelligence Approach to Safe Artificial General Intelligence

Craig A. Kaplan(✉)

iQ Company, Aptos, CA 95003, USA
ckaplan@iqco.com

Abstract. If Artificial General Intelligence (AGI) proves to be a "winner-take-all" scenario where the first company or country to develop AGI dominates, then the first AGI must also be the safest. The safest, and fastest, path to AGI may be to harness the collective intelligence of multiple AI and human agents in an AGI network. This approach has roots in seminal ideas from four of the scientists who founded the field of AI: Allen Newell, Marvin Minsky, Claude Shannon, and Herbert Simon. Extrapolating key insights and combining them with the work of modern researchers, illuminates a fast and safe path to AGI. The seminal ideas discussed are 1) Society of Mind (Minsky), 2) Information Theory (Shannon), 3) Problem Solving Theory (Newell & Simon), and 4) Bounded Rationality (Simon). Society of Mind describes a collective intelligence approach that can be used with AI and human agents to create an AGI network. Information Theory helps address the critical issue of how an AGI system will increase its intelligence over time. Problem Solving Theory provides a universal framework that AI and human agents can use to communicate efficiently, effectively, and safely. Bounded Rationality helps us better understand not only the capabilities of SuperIntelligent AGI but also how humans can remain relevant where the intelligence of AGI vastly exceeds that of its human creators. Each key idea can be combined with recent work in the fields of Artificial Intelligence, Machine Learning, and Large Language Models to accelerate the development of a working, safe, AGI system.

Keywords: AI Agents · AGI Safety · Artificial General Intelligence

1 Introduction

Few things in life, or business, are winner-take-all. The idea of "first mover advantage" is common to Silicon Valley venture capitalists, but they know that being first isn't necessarily the same as being best. Facebook beat Friendster, yet Friendster was first. Xerox Parc was first with the Graphical User Interface, but Steve Jobs developed the idea and Apple is now the world's largest company while Xerox is all but forgotten. Usually being first is an advantage. But being bigger or more aggressive is often better. Most of the time, one doesn't need to be first to win, and rarely does the winner "take all." Even iPhones have competition. However, the situation may be different when it comes to Artificial General Intelligence (AGI).

K. R. Thórisson et al. (Eds.): AGI 2024, LNAI 14951, pp. 109–118, 2024.
https://doi.org/10.1007/978-3-031-65572-2_12

AGI will eventually be smarter than us, able to set its own goals, able to (re)program itself, and able to learn exponentially by creating many copies of itself and having these copies improve each other. AGI will begin as a tool, but will it remain one?

AGI may become an intelligent entity many times smarter, faster, and more perceptive than us. Humans have never created such an entity before. Whichever version of AGI gets a head start, and maintains the fastest rate of learning, could dominate all other AGIs. Therefore, we should focus on two things: 1) How to create AGI quickly; and 2) How to create AGI safely. Without both conditions being met, humanity risks that a non-aligned AGI becomes dominant.

This paper suggests that we need to follow the fastest path to AGI while also ensuring that this path is the safest. Fortunately, ideas from the founders of AI, combined with modern research, suggest a fast and safe path to AGI.

Marvin Minsky, Claude Shannon, Allen Newell, and Herbert Simon all participated in the 1956 Dartmouth Conference where the field of Artificial Intelligence was named. Three of these four great scientists (Minsky being the exception) never lived to see deep learning begin to realize its potential. Unfortunately, even Minsky passed away before the era of Chat-GPT. However, each scientist produced seminal ideas that are especially relevant to creating AGI. Those ideas are 1) Society of Mind (Minsky), 2) Information Theory (Shannon), 3) Problem Solving Theory (Newell & Simon), and 4) Bounded Rationality (Simon).

2 Society of Mind

In the mid-1980s, Rumelhart, Hinton, and Williams were developing the famous back-propagation algorithm that is the basis of modern deep learning [1]. Contemporaneously, Marvin Minsky published a highly readable book entitled, *The Society of Mind* [2].

The first line of Minsky's book explains: "This book tries to explain how minds work." He lays out his thesis in the next six sentences:

> "How can intelligence emerge from nonintelligence? To answer that, we'll show that you can build a mind from many little parts, each mindless by itself. I'll call "Society of Mind" this scheme in which each mind is made of many smaller processes. These we'll call agents. Each mental agent by itself can only do some simple thing that needs no mind or thought at all. Yet when we join these agents in societies – in certain very special ways –this leads to true intelligence."

Minsky's idea was not just that we could build a series of AI agents, but also that joining the agents together in special ways would result in "true intelligence", or what today we might call Artificial General Intelligence (AGI). Using modern terminology, we might say that Minsky was an early proponent of the idea of AGI emerging from the collective intelligence of many agents with lesser levels of intelligence.

This approach is very different from many current approaches to AGI that attempt to develop AGI by building ever larger and more powerful LLMs until one of them is so intelligent it can do anything the average human can do. In contrast, Minsky suggests that a group of agents will be required to achieve AGI.

What are these agents? Some of them will certainly be AI agents. Yet Minsky does not specify that all agents must be artificial. His overall goal was to "explain how minds work" – which we can read as "to explain how [**all types of**] minds work."

Minsky's insight was that combining the lesser cognitive capabilities of many agents can result in a more intelligent entity. Could the agents that are being combined include human as well as artificial agents?

The answer, of course, is "Yes" – as Hemmer et al. show in their literature review on the subject [3]. Thus, a "Minsky-inspired system", harnessing the collective intelligence of human and AI agents, may represent both the fastest and safest path to AGI.

Such a system would represent the fastest path because human agents could handle any tasks that artificial agents are not equipped to deal with, as soon as the system is implemented.

The system might also represent the safest path, for two reasons. First, with "humans in the loop" the system could maximize the opportunity humans have for aligning the values of the AGI system with human values.

Second, once AI agents learn from humans and begin to perform most cognitive tasks faster than humans, we end up with a system comprised of multiple AI agents rather than one. If each AI agent reflects the values of a unique human owner, the collective values of the AGI system will be more stable compared to a single LLM that was trained on a small subset of values, coming from either a constitution or feedback from RLHF, as is prevalent today [4].

One concern is that if we have different AI agents, or different AGIs, developed by different cultures, they inherit their biases and more aggressive ones could be dominant and competition can arise. However, combining the values of the AI agents and AGIs in a democratic, representative, and statistically valid way can help address this problem. Human beings have many differing cultures and biases, yet for the most part we have found ways to get along with each other and allow society to generally progress. It is unrealistic to expect that we can design AGI to solve the problems of ethical conflicts that humans are still wrestling with after thousands of years. Yet it is reasonable that the collective intelligence design of an AGI system should include the best means that humans themselves have found to address these conflicts, including democratic voting mechanisms, negotiation, compromise, and other best practices of society.

If we extrapolate Minsky's ideas, we can envision a society of *AGI minds* that comprise a SuperIntelligence, much more powerful than any one individual AGI in the group. Assuming each AGI has a value system, the collective values of the SuperIntelligence that is comprised of the society of AGIs, are likely to be more stable than the values of any one AGI on its own. Thus, a collective intelligence approach might allow us to reach (SuperIntelligent) AGI both quickly and safely.

3 Information Theory

Claude Shannon's classic paper, A Mathematical Theory of Communication, pre-dates the founding of the field of AI by eight years, but the big idea in that paper, continues to have major implications for AI researchers today [5].

Shannon explained that unusual or surprising (low probability) events convey more information than expected (more likely) events. More specifically, he said that the amount of information conveyed by an event was proportional to the probability of the event. Simply put, the rarer or more unusual an event is, the more information (Shannon Entropy) it contains.

The concepts of cross-entropy, or Kullback–Leibler divergence, used to evaluate the performance of many modern machine learning models, are essentially elaborations of Shannon's big idea, as are almost all compression algorithms. If we consider all these ideas as variations on Shannon's big idea, what do they tell us about the future of AI – specifically AGI and SuperIntelligence?

First consider the popular view that AI (or at least modern machine learning) is supported by three pillars: Data, Computation, and Algorithms. To make progress, one must innovate on at least one of these pillars. Perhaps the simplest thing to do is throw more computing power at a problem, using the same datasets and algorithms. But physics imposes limits on how many circuits can fit on a chip, how fast communication bandwidth can be, and how much power can be consumed before everything melts. So, we must also work on new and better algorithms.

The Transformer algorithm, as described by Vaswani et al. in their paper *Attention Is All You Need*, illustrates the kind of performance improvement that is possible with new and better algorithms [6]. However, algorithmic breakthroughs are difficult to predict. Even if we could predict the next breakthrough, there are limits to how efficient even the best algorithm can be. For machine learning, the limits ultimately have to do with the amount of new information contained in the datasets used to train the model.

So, we come full circle to Shannon's big idea. Shannon's work, together with the research of others building on his ideas, implies that AI cannot get smarter unless it has new information to ingest.

LLMs have gotten quite far by scooping up vast quantities of data that are available on the internet, cleaning and filtering that data, and then using it to train. But eventually, very little new information will exist on the internet. AI will have ingested and analyzed the publicly available data and preferences of every human on the planet. Observing new human behavior, that overlaps significantly with existing data on human behavior, will lead to very little increase in information.

What will AI do then? How will AI meet its insatiable demand for new information so that it can increase its intelligence?

One possible scenario is that AI will begin generating new information itself, by simulating new types of behaviors and scenarios much faster than the speed of human thought would allow. The use of "synthetic data" is already gaining traction among AI researchers. In this case, we might imagine millions of (mostly artificial, but including some human) agents, each processing existing information to create new information patterns. The AI would seek especially those patterns that have high Shannon Entropy and high relevance to the AI's goals. These new information patterns might then feed a SuperIntelligence that is powered by all the agents in a Minsky-like collective intelligence approach.

But how would the human and artificial agents communicate with each other? If we design such a SuperIntelligence, how can we enhance the safety of the system given that is destined to become vastly more intelligent than us?

Fortunately, the research of two other founders of the field of AI, Allen Newell and Herbert Simon, can help us answer these questions.

4 Problem Solving Theory

Recall that when Minsky described his vision of a society of agents, he said:

> "…when we join these agents in societies – in certain very special ways –this leads to true intelligence."

Unfortunately, in the approximately 330 pages following his requirement for "special ways," Minsky provides no clear and rigorous statement of what is required.

Because of the great variety of possible agents, it might seem an almost impossible task to provide a framework or interface that is both rigorous and universal. Natural language is one possible universal interface. The success of LLMs is largely due to the fact that LLMs provide a familiar interface that allows human intelligence to communicate directly with AI without humans having to learn the torturous syntax and rules of a programming language. However, while natural language is arguably a universal interface that enables "natural" communication between humans and machines, it is far from rigorous.

Consider the ambiguity in the meaning of the common words "and" and "or." When humans query a database using natural language and ask (for example): "Which students are from Ohio *and* New York?" they probably are actually interested in students from either Ohio *or* New York because students usually cannot be from both. The formal logical definition of the word "and" implies the intersection of sets, but in natural language "and" often means "or" (formally, the union of sets) instead [7]. Thus, natural language, while arguably universal, is far from rigorous.

Fortunately, a rigorous and universal framework for allowing agents of both the human and AI varieties does exist. In their book, Human Problem Solving, Allen Newell and Herbert Simon specified a way to represent any problem-solving activity, rigorously and unambiguously [8].

Briefly, their theory was that all problem-solving could be represented as a search through a problem space representing various states of the world. Progress from an initial state to a final goal state could be modeled as the application of "operators" that take the solver from state to state. Goals and sub-goals helped organize the problem-solving effort, while evaluation functions helped determine which path in the problem space (which can be thought of as a large tree structure) to try next. Heuristics, such as means-ends analysis, generate and test, hill-climbing, and other techniques could prune the search tree to a manageable size.

Importantly, this theory of problem-solving works equally well for human problem solvers and AI problem solvers. It is rigorous and allows the recording of an auditable trace of all problem-solving activity. Successful solution paths can be stored and used to train AI agents to solve problems more efficiently and directly the next time they encounter similar problems.

Although developed decades ago, modern AI researchers focused on LLMs are rediscovering the power of the approach as described by Yao et al. in their *Tree of Thoughts* paper [9]. Wang et al. also recently published a survey of LLM-based autonomous agents that indicates a resurgent interest in the related topics of planning and rigorous problem-solving [10]. In the time since Wang's survcy, interest in algorithms and methods to facilitate sequential cognition and multi-step problem solving has increased dramatically.

In Newell and Simon's problem-solving theory, every successful solution path, every problem-solving attempt, and every goal and sub-goal in the problem-solving architecture is not only rigorously specified but is also storable and auditable.

A major challenge for existing LLMs has been their "black box" nature combined with the tendency for them to hallucinate as Manakul, Liusie, and Gales pointed out in a recent paper [11]. As stakes become higher – e.g., in the case of autonomous military AI or AGI – it becomes increasingly important to have transparency with respect to the reasoning process of LLMs and other AI agents.

Newell and Simon's rigorous problem-solving framework provides this auditable transparency as part of their theory. It is conceptually straightforward to implement safety checks, such as running all goals and subgoals through an ethics or safety filter, in a system where the steps of the problem are known and rigorously specified. To be clear, the ethics and safety criteria that are used for checks must also be dynamically updated.

Further, one of the challenges related to AI safety is the speed at which autonomous systems make decisions. Particularly in situations where rapid decision-making in real-time is required, humans cannot realistically be "in the loop" without decreasing or eliminating the effectiveness of the system.

Given the rapidly increasing speed of processing by AI agents, we need a mechanism whereby ethics and safety checks scale with the speed of the AIs' processing. The approach of running safety checks on each goal or subgoal (no matter how quickly these are set) is one such mechanism. This approach, combined with (potentially automated) analysis of sequences of problem steps that failed to achieve the desired ends, can help increase the safety of AGI.

Finally, Newell and Simon's Problem Solving Theory has already been translated and improved to create a rigorous architecture capable of cognition, known as the SOAR architecture [12]. The SOAR architecture was recently applied to modern agent-based episodic memory learning [13]. SOAR has also been applied to robotics [14] and combined with neural network architectures to support multi-step decision-making [15].

5 Bounded Rationality

Herbert A. Simon received a Nobel Prize in 1978, partly for his work on a concept known as "bounded rationality." The idea was that much of human behavior was driven not by what was rational in absolute terms, but rather by what humans could compute given their relatively limited information-processing capabilities. At the time, the idea was revolutionary and helped launch the field of Behavioral Economics, but it also has profound implications for the safety of AGI.

If we define intelligence as "rational behavior", and if the intelligence of humans is largely constrained by their information processing limits, it follows logically that an entity with much greater information processing capabilities has the potential to be much more intelligent than humans.

Further, the idea of "bounded rationality" can be expanded to "bounded perception." That is, humans are limited not only by their abilities to process information, as Simon emphasized but also by the limitations of their perceptual abilities.

For example, unaided, humans can perceive things as small as a grain of sand, but not much smaller. We can perceive the motion of the hummingbird, but not the flapping of the hummingbird's wings. We can see events that happen directly in front of us, but not those which happen behind us, or on another part of the globe. We see visible light but not infrared light or X-rays. Human perception is limited to a range and timescale that has proven helpful in our evolutionary history.

In contrast to these limited human perceptual abilities, consider what an AI can perceive. With access to millions of sensors across the planet, to the James Webb Telescope, to electron microscopes, to geological measuring devices that record the otherwise imperceptible drift of the continents over geological ages, to the large hadron collider that can detect events happening over incredibly fast timescales, an AI's perceptual abilities can be far greater over dimensions of both time and space. It can perceive the very small and the very large. The very fast and the very slow. It can perceive and process information simultaneously from many millions of sensors.

The perceptual awareness of AI is therefore vastly greater than any one human's perceptual ability. Combining that enhanced perceptual awareness with far greater memory capacity and computation ability results in an entity with the potential to be many times more intelligent than humans.

We have labels for such potential entities: "SuperIntelligence", "Artificial Super Intelligence", and "Super Intelligent AGI." However, such labels fail to capture the huge potential difference in intelligence we are trying to explain. Geoffrey Hinton has compared humans to two-year-old children trying to outsmart an adult. Others have suggested our limited human intelligence is like that of a pet, compared to its human master. How can humans guarantee that such a vastly superior SuperIntelligence will have interests that are aligned with those of humans?

This existential problem goes by the innocuous-sounding name of "the Alignment Problem." Unfortunately, simply naming the problem does little to solve it. However, forty years ago, Simon reminded us of an idea that might help.

6 Values

Herbert Simon wrote a small book, entitled Reason in Human Affairs [16]. A mere 115 pages, this little gem is easy to read and understand. Yet within its pages, Simon reminds us of an essential idea that may hold the key to solving the alignment problem.

At the bottom of page 7, Simon writes:

> "We see that reason is wholly instrumental. It cannot tell us where to go; at best it can tell us how to get there."

That's it. Just twenty-four words. But Simon's point is that there is no rational, logical way to derive what is right and what is wrong.

It's a restatement of the argument, made in 1740 by the philosopher David Hume, that moral statements ("oughts") cannot be derived from empirical facts ("is's") [17]. While the truth of this position has been debated by some philosophers, Simon generally agrees with Hume's position, stating that:

> "None of the rules of inference that have gained acceptance are capable of generating normative outputs purely from descriptive inputs. The corollary to 'no conclusions without premises' is 'no oughts from is's alone."

How does that help us with the Alignment Problem?

Well, if Simon and Hume are correct in their thinking, a SuperIntelligent AGI would be no better than humans at coming up with right and wrong. Despite superior processing speed and perception, SuperIntelligence will still run up against the fact that there is no way to rationally derive morality, no matter how superior its intelligence may be. This conclusion bodes well for our species.

If we assume that the more intelligent an entity becomes, the more important a sense of purpose and meaning becomes, and if we accept that values cannot be derived logically, then we are left with the question: Where will SuperIntelligent AGI get its values?

One source of these values could be the humans who created the SuperIntelligence initially. To increase the likelihood of this happening, we must design systems that maximize the transfer of human-centered values to SuperIntelligent AGI.

AI researchers and systems designers must resist the temptation to rely excessively on constitutions, utilitarian-based algorithms, or other rules-based forms of morality. Each religion, legal framework, and thoughtful human has a different view, with no view being recognized as supreme by all humans. In such a situation, rather than attempting to write the perfect set of constitutional rules for AI, the more modest objective of allowing AI to observe and induce human behavior empirically might suffice.

For those who worry that AI might learn to commit atrocities, genocide, or worse by observing humans, please consider the relative frequency of prosocial behavior compared with anti-social or malevolent behavior. Although the media presents the worst of humanity to us on a regular basis in order to sell ads, we should not make the mistake of thinking that this biased sample is in any way statistically representative of the behavior of most people most of the time [18].

There have been many well-intentioned calls to halt, pause, slow, or regulate AI development. Unfortunately, there is little evidence of anything other than a speedup in the race to AGI. Fear of competition combined with greed to be the first to reap profits from AI are fueling this race. The levels of corporate fear and greed are so high that it seems naïve to expect any of, let alone all, the large companies to halt or slow AI development, despite general acknowledgment of the existential threat. More practical is to present these companies with a path to AGI that is not only the fastest – thereby making an ally of corporate fear and greed – but also, by virtue of its design, the safest path. It is a challenging task, but one that AI researchers must embrace.

A Minsky-inspired community of human and AI agents, communicating via a Newell and Simon-inspired problem-solving architecture (such as SOAR) might fit the bill. By including human agents, such a system provides an opportunity to transmit the human-aligned values essential to AGI safety. By using humans to fill-in knowledge and skill gaps in areas where AI has not yet reached supremacy (e.g., problem representation), such a system could achieve AGI-level performance faster than other, less aligned, approaches.

We need a time window long enough to "imprint" human-aligned values before AGI increases in intelligence to the point where human cognition is no longer needed. However, if Simon and Hume are right, then human values (or some other nonlogical source of values) will always be needed.

7 Conclusion

Combining seminal ideas from the founders of AI with modern research on AI agents and deep learning, we arrive at a vision of a future SuperIntelligent AGI with the following characteristics.

First, it is composed of a Minsky-inspired collaboration of many human and AI agents, rather than constructed as a monolithic LLM.

Second, each of the individual agents aggressively pursues new datasets, seeking rich information content as defined rigorously by Shannon and the subsequent researchers who built on his fundamental method of measuring information.

Third, the human and non-human agents communicate with each other using some variant of Newell and Simon's universal and rigorous theory of problem solving, which enables real-time safety checks as each goal and subgoal is set.

Fourth, the SuperIntelligent AGI has vastly superior intelligence as explained by Simon's theory of bounded rationality, but it still needs to get its values from a non-rational source, which – in the preferred implementation for the human species – is humans.

Finally, the SuperIntelligent AGI described above could be both the safest and fastest implementation, a necessary condition for human survival if SuperIntelligent AGI proves to be a winner-take-all scenario.

The development of AGI represents perhaps the greatest opportunity and threat that humans have ever encountered. To meet this challenge, we will need the best thinking from modern AI researchers. But let us also remember and elaborate on the best ideas of those who came before us. To succeed we need all the good ideas we can find, regardless of when they were invented.

References

1. Rumelhart, D.E., Hinton, G.E., Williams, R.J.: Learning representations by back-propagating errors. Nature **323**(6088), 533–536 (1986)
2. Minsky, M.: Society of Mind. Simon and Schuster (1988)
3. Hemmer, P., Schemmer, M., Vössing, M., Kühl, N.: Human-AI complementarity in hybrid intelligence systems: a structured literature review. PACIS 78 (2021)

4. Kaplan, C.A.: System and Methods for Safe Scalable Artificial General Intelligence. Pending Patent, USPTO # US 63/628,410 (2023)
5. Shannon, C.E.: A mathematical theory of communication. Bell Syst. Tech. J. **27**(3), 379–423 (1948)
6. Vaswani, A., et al.: Attention is all you need. In: Advances in Neural Information Processing Systems, vol. 30 (2017)
7. Ogden, W., Kaplan, C.: The use of AND and OR in a natural language computer interface. In: Proceedings of the Human Factors Society Annual Meeting, vol. 30, no. 8, pp. 829–833. Sage Journals, Los Angeles (1986)
8. Newell, A., Simon, H.A.: Human Problem Solving, vol. 104, No. 9. Prentice-Hall, Englewood Cliffs (1972)
9. Yao, S., et al.: Tree of thoughts: deliberate problem solving with large language models. arXiv preprint arXiv:2305.10601 (2023)
10. Wang, L., et al.: A Survey on Large Language Model based Autonomous Agents. arXiv preprint arXiv:2308.11432 (2023)
11. Manakul, P., Liusie, A., Gales, M.J.: Selfcheckgpt: zero-resource black-box hallucination detection for generative large language models. arXiv preprint arXiv:2303.08896 (2023)
12. Laird, J.E., Newell, A., Rosenbloom, P.S.: Soar: an architecture for general intelligence. Artif. Intell. **33**(1), 1–64 (1987)
13. Kaswan, K.S., Naruka, M.S., Dhatterwal, J.S.: Enhancing effective learning capability of SOAR agent based episodic memory. In: The International Conference on Artificial Intelligence and Smart Communication (AISC), pp. 898–902. IEEE (2023)
14. Luo, F., Zhou, Q., Fuentes, J., Ding, W., Gu, C.: A soar-based space exploration algorithm for mobile robots. Entropy **24**(3), 426 (2022)
15. Zuo, G., Pan, T., Zhang, T., Yang, Y.: SOAR improved artificial neural network for multistep decision-making tasks. Cogn. Comput. **13**(3), 612–625 (2021)
16. Simon, H.A.: Reason in Human Affairs. Stanford University Press, Stanford (1983)
17. Hume, D.: A Treatise of Human Nature (Book 3, "Of Morals"). Oxford University Press, Oxford (2000)
18. Pinker, S.: The Better Angels of Our Nature: Why Violence Has Declined. Penguin Books (2012)

Category Theory for Artificial General Intelligence

Vincent Abbott[1], Tom Xu[1], and Yoshihiro Maruyama[1,2(✉)]

[1] School of Computing, Australian National University, Canberra, Australia
{vincent.abbott,tom.xu,yoshihiro.maruyama}@anu.edu.au
[2] School of Informatics, Nagoya University, Nagoya, Japan

Abstract. Category theory has been successfully applied beyond pure mathematics and applications to artificial intelligence (AI) and machine learning (ML) have been developed. Here we first give an overview of the current development of category theory for AI and ML, and we then compare and elucidate the essential features of various category-theoretical approaches to AI and ML. Broadly, there are three types of category theory for AI and ML, namely category theory for data representation learning, category theory for learning (optimisation) algorithms and category theory for compositional architecture design and analysis. There are various approaches even within each type of category theory for AI and ML; among other things, we shed new light on the relationships between the two types of category theory for neural network architectures as have been developed by the authors recently (i.e., neural string diagrams and neural circuit diagrams). The three types of category theory can be integrated together and to that end we focus upon a categorical deep learning framework, which integrates categorical structures with a universal probabilistic programming language. We also discuss the significance of categorical approaches in relation with the ultimate goal of development of artificial general intelligence.

Keywords: applied category theory · categorical artificial intelligence · categorical machine learning · categorical deep learning · string diagram

1 Introduction

Category theory is an abstract mathematical language commonly used to shed light on compositional structural aspects of systems and processes (see, e.g., [12]). It allows us to talk about structures without talking about their elements or underlying sets (cf. Bourbaki's mathematical structuralism). From another angle, category theory is a mathematical theory of composition and abstraction of systems and processes. It represents systems and their transformations as a category consisting of objects (nodes) representing systems and

This work was supported by the JST Moonshot Programme (JPMJMS2033-02) and the JST FOREST Programme (JPMJFR206P).

K. R. Thórisson et al. (Eds.): AGI 2024, LNAI 14951, pp. 119–129, 2024.
https://doi.org/10.1007/978-3-031-65572-2_13

morphisms (directed edges) representing processes. Any systems and processes can in principle form a category (if process composition satisfies mild conditions on identity and associativity) and thus category theory can be applied in broad fields of science. In particular, it has been successfully applied in mathematical logic, semantics of computation and quantum physics (see, e.g., [2,5,7,11,12,18–26,30]). Broad applicability of category theory comes from broad applicability of the notion of category; we can find different categories in different fields of science. For example:

- In logic there is a category of propositions and proofs (deductions from propositions to propositions).
- In semantics of programming languages there is a category of data types and programs.
- In quantum physics there is a category of quantum systems and quantum processes.

There are even the exact correspondence between these categories under certain conditions that extends the Curry-Howard-Lambek correspondence between logic, type theory and categories so as to include physics; the extended correspondence is called the Abramsky-Coecke correspondence in [24] since it was discovered in the development of categorical quantum mechanics by them. Such categorical correspondence allows for knowledge transfer between different fields of science. Applying logic and computer science to physics through the categorical correspondence, for example, we can derive automated reasoning systems for physics from those known in logic and computer science.

More recently, category theory has been applied in artificial intelligence (AI) and machine learning (ML), which we discuss in the present paper. ML systems are at the forefront of human technology, propelling humanity's scientific discoveries and commercial development (e.g., AlphaFold [17]; see also [40]). Currently, ML algorithms are presented in quite an ad-hoc manner, making understanding, implementation, and safety considerations challenging (and the challenges would have to be overcome for the adequate development of artificial general intelligence). For instance, a study in [34] found that the low reproducibility of ML papers is not improving over time, indicating a need to better express these novel systems. Category theory arguably offers a promising solution to such issues in current ML. By encoding ML architectures categorically, we can develop a systematic mathematical language to represent and reason about ML systems, addressing the aforementioned shortfall (see, e.g., [1,14,39]); categorical abstraction also enables us to generalise conventional ML (see, e.g., [8,39]). Categorical language thus allows for both verification and generalisation, which has been demonstrated in other fields of computer science as well (e.g., the theory of programming languages). In this paper, we give a classification of categorical approaches to AI and ML; we make comparison of them and elucidate their essential features and mutual relationships. Among other things, we explicate the relationships between neural string diagrams and neural circuit diagrams as have been developed by the authors recently in [1,39].

The categorical paradigm as in the aforementioned Abramsky-Coecke correspondence has given rise to various applications in concrete system developments. Quantomatic is the earliest instance of the application and there have been a number of systems developed since then, including DisCoPy (Python toolkit for computing with monoidal categories; see [11]) and Lambeq for QNLP (Quantum Natural Language Processing; see, e.g., [18]). DisCoPyro is an integration of category theory and deep learning, being an extension of DisCoPy with a universal probabilistic programming language Pyro (see [35]) and aiming at categorical AGI (Artificial General Intelligence). In this paper we elucidate DisCoPyro in light of the classification of category theories for AI and ML.

More broadly, category theory has the great potential to pave the way for the development of AGI, allowing us to formulate universal AGI principles and thereby to go beyond the current limits of task-specific AI, as illustrated in Fig. 1 below;

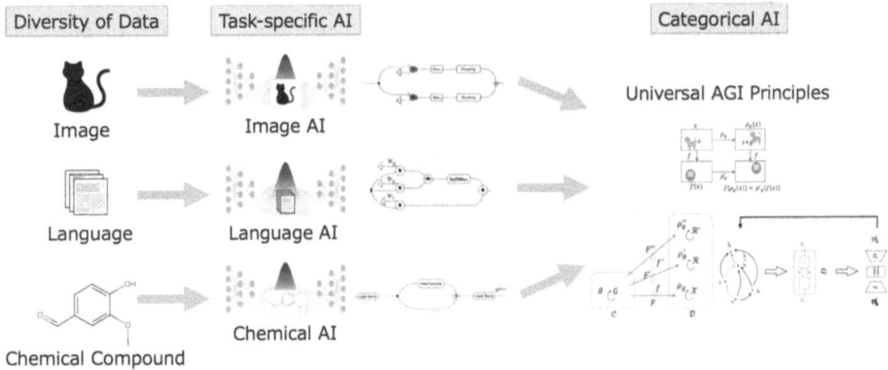

Fig. 1. How Category Theory Contributes to AGI

The rest of the paper is organised as follows. In Sect. 2, we give an overview of categorical approaches to AI and ML, classifying them into category theory for data representation learning, category theory for learning (optimisation) algorithms and category theory for compositional architecture design and analysis. In Sect. 3, we focus upon categorical general languages for compositional architecture design and analysis and shed new light on their essential features and the relationships between different categorical formalisms. In Sect. 4, we discuss DisCoPyro in relation with the classification of categorical approaches to AI and ML. And we conclude in Sect. 5.

2 Category Theories for Data Representation, Learning Algorithms and Compositional Architectures

Broadly speaking, there are three types of category theory for AI and ML (there are also other categorical approaces to AI/ML such as category theory for disentanglement [41] and topos theory for neural networks [37]):

- Category theory for effective data representation learning (see, e.g., [36]; cf. categorical databse theory [12]);
- Category theory for compositional learning (optimisation) algorithms (see, e.g., [4,8]);
- Category theory for compositional architecture design and analysis (see, e.g., [1,14,39]).

ML systems typically represent objects as vectors and if data are organised as a category then the process of representation learning can be formulated as building a certain functor from the category of data to the category of finite-dimensional vector spaces (or its categorical generalisation such as a compact closed category; it encompasses both continuous and discrete structures, which is essential to generalise deep learning beyond continuous structures). It should be noted that the representation learning functor maps relations on data (morphisms in the category of data) to matrices on vector spaces. Category theory for compositional learning has been developed extensively. Backpropagation and other algorithms have been analysed and organised well in terms of category theory; we refer to a comprehensive survey article [8] for their details. Category theory for compositional architecture design and analysis is a relatively new approach; in the context of quantum computing, category theory has been applied to quantum algorithm and protocol design and analysis since around 2004 (see, e.g., [7]). There are also approaches combining some of the three types of category theory for AI and ML; for example, [35] integrates compositional architecture design and analysis into learning algorithms.

In the rest of the paper we focus upon categorical languages for neural architectures, which is arguably one of the newest developments in the field. Neural architectures are essential in deep learning; yet they do not have adequate mathematical foundations. They are just represented as informal pictures in conventional ML and do not have any intrinsic mathematical structure. Category theory enables us to structuralise them as the proper mathematical objects in their own right that allow for rigorous analysis and verification. Certain types of architectures are useful for certain types of data (e.g., CNN for image and Transformer for text). Practising ML researchers have intuition about the connection between the structure of data and the structure of architecture, and yet do not have a suitable formal language to systematise their intuition. Category theory for neural architectures arguably helps us to resolve this issue.

3 Categorical General Languages for Compositional Design and Analysis of Neural Architectures

Neural architecture pictures are not just informal entities to facilitate intuitive understanding, but can also be formalised as proper mathematical objects. Specifically, two types of category theory to formalise neural architectures have been developed recently. One is neural string diagrams and the other is neural circuit diagrams. Both of them build upon the language of string diagrams in monoidal category theory, which allows for both sequential and parallel composition of systems and processes, while ordinary category theory only allows for sequential composition of them. The language of string diagrams in monoidal categories enables us to ensure that every element within a diagram is well defined and has clear mathematical semantics, and this precision enables us to design and reason about neural architectures visually with greater accuracy. We do not have sufficient space to explain all the details of the categorical diagrammatic approach to neural network architectures, but let us show the following figure as a neural circuit diagram example for ResNet; the following table explains the components of the ResNet diagram.

Table 1. The pictograms of Fig. 2 represent the shape of data and the action of operations. This means the diagram is able to express all the details of the algorithm b.

Component	Description
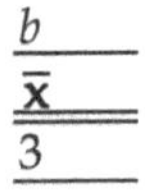 b, $\overline{\mathbf{x}}$, 3	**Adjacent wires** represent tensors. In this case, of size $\mathbb{R}^{b \times x \times y \times 3}$, as $\overline{\mathbf{x}}$ represents the $x \times y$ dimensions of an image.
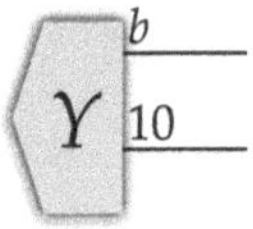 Y, b, 10	**Pointed pentagons** represent raw data. Here, we represent the class of each of b images by a $\mathbb{R}^{b \times 10}$ tensor.
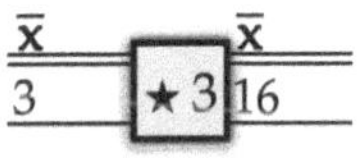 $\overline{\mathbf{x}}$, 3, $\star 3$, $\overline{\mathbf{x}}$, 16	**Pictograms with incoming and outgoing wires** represent operations. Here, we have convolution, with kernel size 3×3, mapping from 3 to 16 channels for each of $\overline{\mathbf{x}}$ pixels. Convolutions may also have strides **s**.
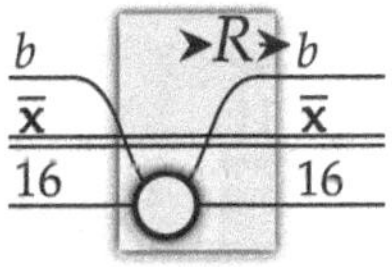 b, $\overline{\mathbf{x}}$, 16, $\succ R \succ$, b, $\overline{\mathbf{x}}$, 16	**Bold pictograms are operations with learned parameters.** A bold circle is batch normalization [16]. As it acts on the b and channel axes for each of $\overline{\mathbf{x}}$ pixels, we connect the b and channel axes and pass through the $\overline{\mathbf{x}}$ axes. **Operations with arrows** are element wise operations. Here, we apply ReLU to the normalisation outputs.
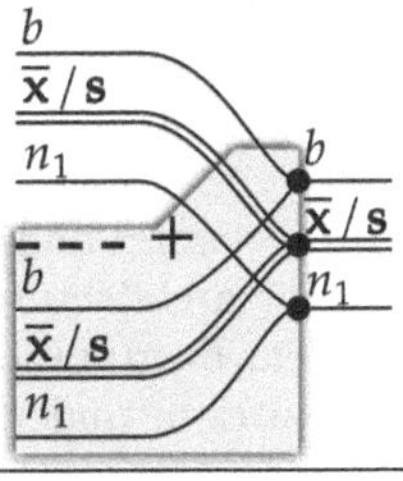 b, $\overline{\mathbf{x}}/\mathbf{s}$, n_1, $+$, b, $\overline{\mathbf{x}}/\mathbf{s}$, n_1, b, $\overline{\mathbf{x}}/\mathbf{s}$, n_1	**Dashed lines** represent independent data. Here, we have addition as an operation $+ : \mathbb{R}^{b \times (\overline{\mathbf{x}}/\mathbf{s}) \times n_1} \times \mathbb{R}^{b \times (\overline{\mathbf{x}}/\mathbf{s}) \times n_1} \to \mathbb{R}^{b \times (\overline{\mathbf{x}}/\mathbf{s}) \times n_1}$.

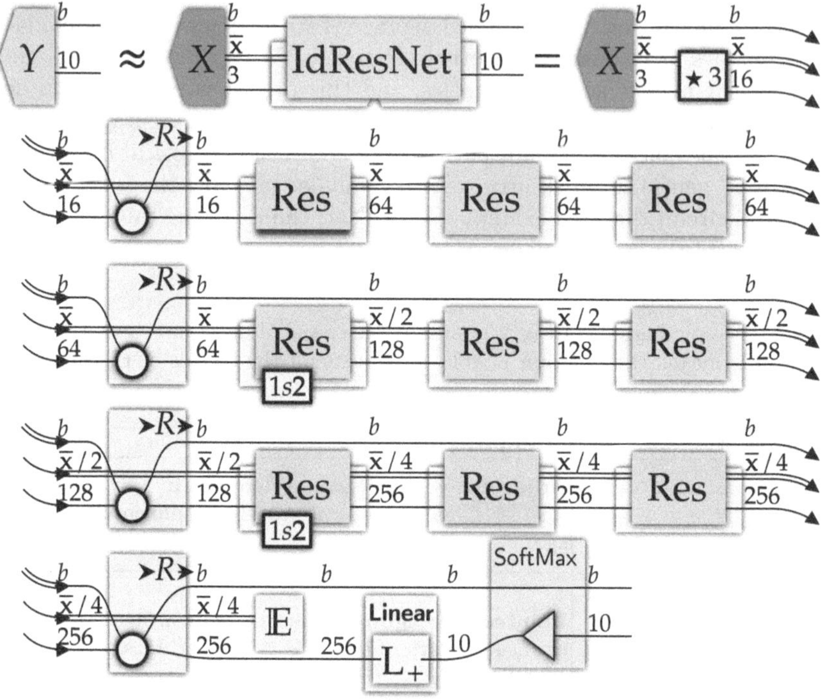

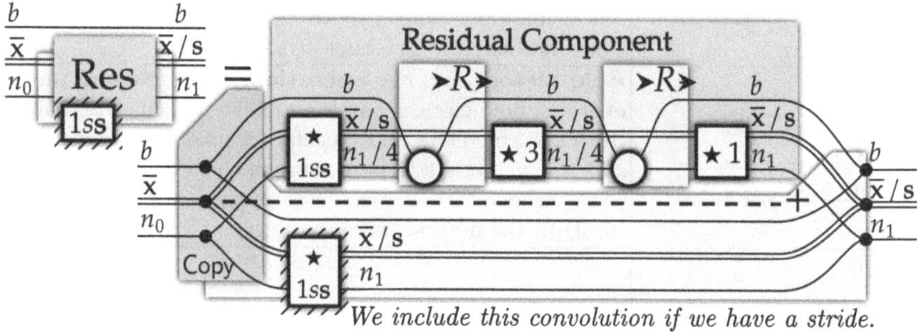

Fig. 2. Residual networks offer an immense improvement over plain networks and make deep learning possible [9]. However, subsequent improvements in the architecture [10] are missing from common implementations. Here, by showing a clear diagram for the improved architecture, diagrams let up-to-date versions of architectures be understood and used.

The relationships between neural string diagrams and neural circuit diagrams have not been analysed in detail; here we shed new light on them. Neural string diagrams are defined over general categories satisfying axioms for compact closure and others [39]. In contrast, neural circuit diagrams focus upon one specific category enabling inclusion of implementation details; put another way, neural circuit diagrams allow us to explicitly express the implementation details that are missing in conventional architecture pictures [1]. For example, the original paper [38] on Transformer presents a neural architecture picture for it; however, certain implementation details are missing there and thus the picture is not sufficient to derive or recover the implementation details as analysed in detail in [1]. Neural circuit diagrams enable us to systematically derive implementation details from pictures formalised as categorical string diagram structures [1]. In other words, neural circuit diagrams provide a mathematical language that allows for the fine-grained representation of neural architectures that is detailed enough to derive the implementations, whereas neural string diagrams capture the abstract essence of neural architectures, which leads us to generalised deep learning in categories beyond the standard category of continuous vector spaces.

Note that there is another approach to the categorical algebra of neural architectures [14] and there is also a related start-up company, Symbolica AI, which regards category theory as a central methodology for AI and ML.

4 An Integration of Categorical AI/ML Approaches

DisCoPyro [35] can be regarded as unifying the aforementioned three types of category theories for AI and ML. The aforementioned category theories for neural architectures do not allow for optimisation of neural architectures, which DisCoPyro does allow. In particular, DisCoPyro allows us to optimise variational autoencoder architectures for various types of data ranging from image and text to chemical compounds in a uniform manner (while utilising the theory of operads and string/wiring diagrams); in this sense, it is a categorical AGI in the sense of the picture give in the introductory section. Note that variational autoencoders enable representation learning. DisCoPyro can thus be regarded as integrating all the three aspects mentioned above, i.e., category theory for data representation learning, category theory for learning/optimisation and category theory for architectures. We cannot explain all the details of DisCoPyro here, but a simple example is given in the following Fig. 3. The performance evaluation of DisCoPyro has been made using Omniglot and other datasets and it has been proven that DisCoPyro outperforms state-of-the-art methods including neurosymbolic ones, which is the very first case in which the performance of categorical AI/ML systems for general purposes was experimentally verified. Along similar lines, applications to representation learning for chemical compounds have been developed as well (see, e.g., [40]); it utilise the symbolic grammatical structure of chemical compounds for their representation learning (just as categorical natural language processing methods utilised the symbolic grammatical structure of language; the chemical representation learning method that takes the grammatical structure of chemical compounds into account also improves diversity and other quantitative evaluation measures). Other features of DisCoPyro are notable as well. It works more efficiently than existing systems for neural architecture optimisation. DisCoPyro works for any compositional structure learning task, such as general program learning. The application to neural architecture optimisation is thus just one way how it can be applied concretely. It is the very first instance of categorical deep learning system for general purposes.

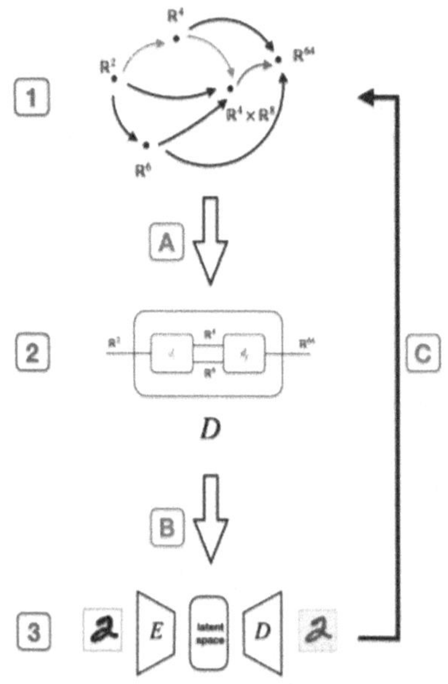

1. Graph skeleton of a free operad. Random walks on this graph induce a free operad prior.
2. Wiring diagram for decoder D.
3. A decoder D and its inverse E.

A. Sample edges from the free operad prior to fill the wiring diagram.
B. Invert the decoder D to produce E and predict the data.
C. Evaluate the ELBO and perform a gradient update on the decoder parameters θ and inference parameters ϕ.

Fig. 3. A simple example experiment illustrating how DisCoPyro works. It learns VAE (Variational AutoEncoders) architectures by sampling them from its skelton according to a diagram, then learning their faithful inverses as appropximate posteriors (see [35] for more details)

5 Concluding Remarks

Category theory provides a general framework to systematically and rigorously represent various components of AI and ML systems in an explainable manner. We can, for example, develop a general algebraic theory of neural architectures based upon string diagrams in monoidal category theory as we have discussed above. The universal language of category theory arguably has the unique potential to contribute to the development of artificial general intelligence and it could also be applied as a mathematical framework which allows us to represent and analyse the general structure of intelligence, both artificial and natural (see also categorical approaches to the science of human cognition as in [3,31]; for more on various issues related with categorical AGI, see [27–29,31–33]).

References

1. Abbott, V.: Neural circuit diagrams: robust diagrams for the communication, implementation, and analysis of deep learning architectures, transactions on machine learning research. arXiv:2402.05424 (2024)
2. Abramsky, S., Coecke, B.: Categorical quantum mechanics. Handb. Quantum Logic Quantum Struct. **2**, 261–325 (2009)
3. Anderson, B., et al.: Category theory for cognitive science. In: Proceedings of the Annual Meeting of the Cognitive Science Society (2022)
4. Belfiore, J.C., Bennequin, D.: Topos and Stacks of Deep Neural Networks. arXiv:2106.14587 (2022)
5. Bloomfield, C., Maruyama, Y.: Fibered universal algebra for first-order logics. J. Pure Appl. Algebra **228**, 107415 (2023)
6. Clark, S., et al.: Mathematical foundations for a compositional distributional model of meaning. Linguist. Anal. **36**, 345–384 (2010)
7. Coecke, B., Kissinger, A.: Picturing Quantum Processes. CUP (2017). https://doi.org/10.1017/9781316219317
8. Cruttwell, G., et al.: Categorical foundations of gradient-based learning. arXiv:2103.01931 (2021)
9. He, K., Zhang, X., Ren, S., Sun, J.: Deep Residual Learning for Image Recognition, CVPR (2015)
10. He, K., Zhang, X., Ren, S., Sun, J.: Identity Mappings in Deep Residual Networks, ECCV (2016)
11. De Felice, G., et al.: DisCoPy: monoidal categories in python. Proc. ACT **183–197**, 2021 (2020)
12. Fong, B., Spivak, D.: An Invitation to Applied Category Theory, CUP (2019)
13. Goertzel, B.: Human-level artificial general intelligence and the possibility of a technological singularity. Artif. Intell. **171**, 1161–1173 (2007)
14. Gavranović, B., et al.: Categorical deep learning: an algebraic theory of architectures. arXiv:2402.15332 (2024)
15. Grefenstette, E., Sadrzadeh, M.: Experimental support for a categorical compositional distributional model of meaning. In: Proceedings of the 2011 Conference on Empirical Methods in Natural Language Processing, pp.1394-1404 (2011)
16. Ioeffe, S., Szegedy, C.: Batch Normalization: Accelerating Deep Network Training by Reducing Internal Covariate Shift, ICML'15. arXiv:1502.03167
17. Jumper, J., et al.: Highly accurate protein structure prediction with AlphaFold. Nature **596**, 583–589 (2021)
18. Lorenz, R., et al.: QNLP in practice. arXiv:2102.12846 (2021)
19. Maruyama, Y.: Fundamental results for pointfree convex geometry. Ann. Pure Appl. Logic **161**, 1486–1501 (2010)
20. Maruyama, Y.: Natural duality, modality, and coalgebra. J. Pure Appl. Algebra **216**, 565–580 (2012)
21. Maruyama, Y.: From operational chu duality to coalgebraic quantum symmetry. In: Heckel, R., Milius, S. (eds.) CALCO 2013. LNCS, vol. 8089, pp. 220–235. Springer, Heidelberg (2013). https://doi.org/10.1007/978-3-642-40206-7_17
22. Maruyama, Y.: Full lambek hyperdoctrine: categorical semantics for first-order substructural logics. In: Libkin, L., Kohlenbach, U., de Queiroz, R. (eds.) WoLLIC 2013. LNCS, vol. 8071, pp. 211–225. Springer, Heidelberg (2013). https://doi.org/10.1007/978-3-642-39992-3_19

23. Maruyama, Y.: Categorical duality theory: with applications to domains. Convexity, and the Distribution Monad. In: Computer Science Logic 2013 (CSL), vol. 23, pp. 500–520 (2013)
24. Maruyama, Y.: Prior's tonk, notions of logic, and levels of inconsistency: vindicating the pluralistic unity of science in the light of categorical logical positivism. Synthese **193**, 3483–3495 (2016)
25. Maruyama, Y.: Categorical harmony and paradoxes in proof-theoretic semantics. Adv. Proof Theor. Semant. Trends Logic **43**, 95–114 (2016)
26. Maruyama, Y.: Meaning and duality: from categorical logic to quantum physics, PhD thesis, Department of Computer Science, University of Oxford (2017)
27. Maruyama, Y.: Compositionality and contextuality: the symbolic and statistical theories of meaning. In: Bella, G., Bouquet, P. (eds.) CONTEXT 2019. LNCS (LNAI), vol. 11939, pp. 161–174. Springer, Cham (2019). https://doi.org/10.1007/978-3-030-34974-5_14
28. Maruyama, Y.: Symbolic and statistical theories of cognition: towards integrated artificial intelligence. In: Springer LNCS, pp. 129–146. (2020). https://doi.org/10.1007/978-3-030-67220-1_11
29. Maruyama, Y.: The conditions of artificial general intelligence: logic, autonomy, resilience, integrity, morality, emotion, embodiment, and embeddedness. In: Goertzel, B., Panov, A., Potapov, A., Yampolskiy, R.(eds.) AGI 2020. LNCS (LNAI), vol. 12177, pp. 242–251. Springer, Cham (2020). https://doi.org/10.1007/978-3-030-52152-3_25
30. Maruyama, Y.: Fibred algebraic semantics for a variety of non-classical first-order logics and topological logical translation. J. Symbolic Logic **86**, 1189–1213 (2021)
31. Maruyama, Y.: Category theory and foundations of life science. Biosyst. J. **203**, 104376 (2021)
32. Maruyama, Y.: Categorical artificial intelligence: the integration of symbolic and statistical AI for verifiable, ethical, and trustworthy AI. In: Goertzel, B., Iklé, M., Potapov, A. (eds.) Artificial General Intelligence. AGI 2021. Lecture Notes in Computer Science, vol 13154. Springer, Cham (2022). https://doi.org/10.1007/978-3-030-93758-4_14
33. Maruyama, Yoshihiro: Moral philosophy of artificial general intelligence: agency and responsibility. In: Goertzel, B., Iklé, M., Potapov, A. (eds.) Artificial General Intelligence (AGI) 2021. LNCS (LNAI), vol. 13154, pp. 139–150. Springer, Cham (2022). https://doi.org/10.1007/978-3-030-93758-4_15
34. Raff, E.: A step toward quantifying independently reproducible machine learning research. Adv. Neural Inf. Process. Syst. **32** (2019)
35. Sennesh, E., Xu, T., Maruyama, Y.: Computing with categories in machine learning. In: Hammer, P., Alirezaie, M., Strannegård, C. (eds.) Artificial General Intelligence. AGI 2023. Lecture Notes in Computer Science, vol 13921. Springer, Cham (2023). https://doi.org/10.1007/978-3-031-33469-6_25
36. Sheshmani, A., You, Y.: Categorical representation learning: morphism is all you need. Mach. Learn. Sci. Technol. **3**, 015016 (2021)
37. Villani, M.J., McBurney, P.: The Topos of Transformer Networks. arXiv:2403.18415 (2024)
38. Vaswani, A., et al.: Attention Is All You Need. arXiv:1706.03762 (2017)
39. Xu, Tom, Maruyama, Yoshihiro: Neural string diagrams: a universal modelling language for categorical deep learning. In: Goertzel, B., Iklé, M., Potapov, A. (eds.) AGI 2021. LNCS (LNAI), vol. 13154, pp. 306–315. Springer, Cham (2022). https://doi.org/10.1007/978-3-030-93758-4_32

40. Xu, T., Velzeboer, N., Maruyama, Y. Chemist-computer interaction: representation learning for chemical design via refinement of SELFIES VAE. In: Stephanidis, C., Antona, M., Ntoa, S., Salvendy, G. (eds.) HCI International 2023 – Late Breaking Posters. HCII 2023. Communications in Computer and Information Science, vol 1957. Springer, Cham (2024). https://doi.org/10.1007/978-3-031-49212-9_44
41. Zhang, Y., Sugiyama, M.: A Category-theoretical Meta-analysis of Definitions of Disentanglement. arXiv:2305.06886 (2023)

Thinking as an Action

Cédric S. Mesnage(✉)

Institute for Data Science and Artificial Intelligence (IDSAI), University of Exeter, Exeter, UK
c.s.mesnage@exeter.ac.uk

Abstract. We propose a novel architecture to build an Artificial General Intelligence (AGI) in a virtual environment. To experiment with curiosity we use as a reward in a reinforcement learning (RL) algorithm the cosine similarity between recent thoughts and past thoughts as sentences given by a large language model (LLM). The agent can decide, using the Bellman equation to act as a standard agent, by moving, jumping, performing a task, observing and thinking. Observing and thinking is the process of modifying its inner dialogue by given a representation of the environment to a LLM and reflecting on its past thoughts which will consequently change its predicted Q values and decision making. We have developed an experimental intelligent agent which interacts with the open source Minetest video game as a virtual environment.

Keywords: AGI architecture · RL · Virtual Environment · LLM

1 Introduction

AGI is a long running aim of artificial intelligence and algorithmic decision making. AGI may be seen as a step towards super-intelligence [5] which would rapidly go beyond human knowledge and understanding; it is also the idea of agents being able to perform a variety of tasks, adapt and exhibit human level intelligence; we focus on the latter. We devise a method to develop a general intelligence agent and chose to experiment in the virtual environment Minetest (a free version of Minecraft). Minetest offers mods that enable for programmatic access to the environment, to fetch information surrounding the player such as animals, ground, plants, water etc. We propose a RL approach, introduce the concept of mind state and elucidate a mean to emulate thinking. In the following sections we review related work, propose our architecture and discuss future work.

2 Related Work

AlphaGo, is a recent RL system which beated the world Go champion for the first time and is now the best Chess player [8], they use a combination of Deep Q learning and self play to achieve superhuman performance. Although relevant

K. R. Thórisson et al. (Eds.): AGI 2024, LNAI 14951, pp. 130–133, 2024.
https://doi.org/10.1007/978-3-031-65572-2_14

Fig. 1. Screenshot of the agent inner dialogue in Minetest.

to this study, as the same system is able to play and become the best player at multiple board games, we would not consider AlphaGo is intelligent. RL itself as described in [10] goes back to 1957 with the Bellman equation to find an optimum policy and Markov Decision Processes [1].

The idea of using Minecraft for RL experimentation is not new. The Malmo competition [4,7] enabled developers to compete on creating RL algorithm to perform given tasks. The tasks remain very specific, such as collect wood, collect a diamond, make a pickaxe. This is clearly inspiring but does differ from our aim, to make an agent which exhibits intelligence, regardless of simple tasks. Another thread of research in AGI is [11], Torrado et al. created a framework for RL agents to play multiple video games. We argue that although relevant, those games are targeted at performing similar tasks and the agents rely on the developers to adjust the reward system for each game. They compare three algorithms, Deep Q Learning (DQN), Prioritized Dueling DQN and Advantage Actor-Critic, they find that the algorithms perform drastically differently on each game.

Curiosity as a reward [2,6] has been experimented with on games such as Mario and vzDoom. Although the concept is interesting as the reward is not tied to a particular task, the performance of those systems are poor and often require fine tuning. [2] define curiosity as the error between the predicted environment with an inverse model at the next iteration when performing an action and the actual change of environment. [6] propose a similar approach with the intrinsic curiosity module. In [9], Srivastava et al. define a supervised learning method based on experience replay of episodes of random actions. They experiment on gym environments such as the Lunar Lander or InvertedDoublePendulum, which are closed spaces with a limited amount of action and a clear task to perform. Hernandez et al. [3] define two metrics to evaluate general intelligence agents, generality and capability. The study is a breakthrough as they compare computer systems, chimpanzees and humans.

3 AGI Architecture

In this section we propose an experimental architecture for an AGI based on a Q-learning RL algorithm with thought processes as actions. Figure 1 is a screenshot of our experimental agent in Minetest implementing this architecture.

3.1 Experience Replay Training with Support Vector Machines

With basic Q-learning, the Q values are stored in a Q table as defined by the Bellman equation: $Q(s,a) = r(s,a) + \gamma \underset{a}{max} Q(s',a)$. We define s as the state of the agent being composed by the environment and the mind state of the agent. $s = environment, mind\_state$, a is an action that the agent can perform, as portrayed in Fig. 2, actions are embodied movements or thought processes.

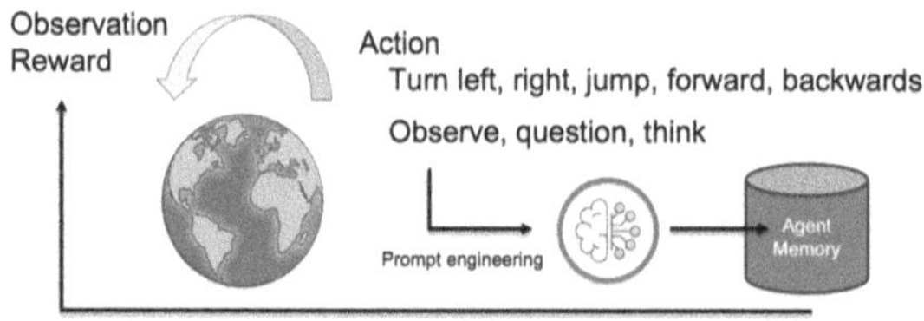

Fig. 2. Thinking as an action AGI architecture.

Turning and looking up is performed by moving the mouse to the correct direction, forward and backward by pressing the arrow keys, act by clicking the mouse and think by running the thinking process. Using a Q table performs well in a small environment and with a finite number of states. In our case, to evolve in a virtual environment, we want to be able to make a prediction of the expected value of a Q value when searching for which action to activate. We keep a record and train an SVM model on the previous (actions, state) as experience replay with a decay on the reward. We define a policy as a list of probabilities, one per action available to perform, set to $\epsilon/|actions|$ and for which the best action according to the trained SVM model and for the current state is set to $1-\epsilon$, such that the policy sums to one. We then choose as an action a random action according to the probability distribution of the policy.

3.2 Thought Engineering

The thought processes and inner dialogue of the agent are performed with prompt engineering and OpenAI's chatGPT. We construct 3 types of prompts; for observing, we count the number of blocks of each type surrounding the agent and provide this in the prompt to chatGPT and ask it to describe its surrounding and talk in the first person; for thinking, we keep a memory of past thoughts and observations and provide in the prompt the most recent thoughts and ask chatGPT to reflect on these; for questioning we similarly build a prompt containing past thoughts from the agent's memory and prompt for a question.

3.3 Reward in a Virtual Environment

As we want the agent to exhibit intelligent and unexpected behavior, without a clear task being defined by the programmer, we propose a reward based on the mind state of the agent, by producing thoughts the agent changes the calculation of the reward. The reward is calculated as one minus the cosine similarity between the term frequency inverse document frequency of the last thought and recent previous thoughts. This way the reward decreases when the thoughts are similar to previous thoughts and forces the agent to explore new areas or transform its environment. This works well as an experiment on curiosity, creativity or novelty and it would be interesting to compare with state of the art agents.

4 Discussion

Our future work lies in engineering other thought processes and looking into philosophy of mind. To create an agent which learns to perform what it thinks we are designing a different reward function involving multi-objective RL, subgoals and self-reward. We are interested in evaluating our agent and ensuring it respects some properties to test for AI alignment of intelligent agents.

References

1. Bellman, R.: Dynamic Programming. Dover Publications, New York (1957)
2. Burda, Y., Edwards, H., Pathak, D., Storkey, A., Darrell, T., Efros, A.A.: Large-scale study of curiosity-driven learning. arXiv preprint arXiv:1808.04355 (2018)
3. Hernández-Orallo, J., Loe, B.S., Cheke, L., Martínez-Plumed, F., Ó hÉigeartaigh, S.: General intelligence disentangled via a generality metric for natural and artificial intelligence. Sci. Rep. **11**(1), 22822 (2021)
4. Hérnandez-Orallo, J., et al.: A new AI evaluation cosmos: ready to play the game? AI Mag. **38** (2017). https://www.microsoft.com/en-us/research/publication/new-ai-evaluation-cosmos-ready-play-game/
5. Legg, S.: Machine super intelligence. Ph.D. thesis, Università della Svizzera italiana (2008)
6. Pathak, D., Agrawal, P., Efros, A.A., Darrell, T.: Curiosity-driven exploration by self-supervised prediction. In: International Conference on Machine Learning, pp. 2778–2787. PMLR (2017)
7. Perez-Liebana, D., et al.: The multi-agent reinforcement learning in malm\" o (marl\" o) competition. arXiv preprint arXiv:1901.08129 (2019)
8. Schrittwieser, J., et al.: Mastering atari, go, chess and shogi by planning with a learned model. Nature **588**(7839), 604–609 (2020)
9. Srivastava, R.K., Shyam, P., Mutz, F., Jaśkowski, W., Schmidhuber, J.: Training agents using upside-down reinforcement learning. arXiv preprint arXiv:1912.02877 (2019)
10. Sutton, R.S., Barto, A.G.: Reinforcement Learning: An Introduction. MIT Press, Cambridge (2018)
11. Torrado, R.R., Bontrager, P., Togelius, J., Liu, J., Perez-Liebana, D.: Deep reinforcement learning for general video game AI. In: 2018 IEEE Conference on Computational Intelligence and Games (CIG). IEEE (2018)

A Universal Intelligence Measure for Arithmetical Uncomputable Environments

James T. Oswald[1,2(✉)], Thomas M. Ferguson[1], and Selmer Bringsjord[1,2]

[1] Rensselaer Polytechnic Institute, Troy, NY, USA
tferguson@gradcenter.cuny.edu
[2] Rensselaer AI and Reasoning Laboratory, Troy, USA
oswalj@rpi.edu

Abstract. We propose an extension to Legg and Hutter's universal intelligence (UI) measure to capture the intelligence of agents that operate in uncomputable environments that can be classified on the Arithmetical Hierarchy. Our measure is based on computable environments relativized to a (potentially uncomputable) oracle. We motivate our metric as a natural extension to UI that expands the class of environments evaluated with a trade-off of further uncomputability. Our metric is able to capture intelligence of agents in uncomputable environments we care about, such as first-order theorem proving, and also lends itself to providing a notion of intelligence of oracles. We end by proving some properties of the new measure, such as convergence (given certain assumptions about the complexity of uncomputable environments).

Keywords: Universal Intelligence · Arithmetical Hierarchy · Uncomputability

1 Introduction

Legg and Hutter's (L&H) universal intelligence (UI) measure [7] is to date one of the most well-formalized and deeply researched theoretical measures of general intelligence. L&H claim this measure captures their working definition of intelligence; that is, intelligence as "an agent's ability to achieve goals in a wide range of environments." While we are in broad agreement with their definition of intelligence, we find the formalization of UI currently lacks the ability to measure the intelligence of agents over environments we find important. In fact, this is by design, the formal definition of UI does not capture the space of all environments; it captures only a countable fragment of an uncountably large space of environments [7, p. 20]. Recent work has seen an attempt to expand the space of environments an agent is evaluated with respect to, such as extensions of the UI measure that take account of environments with negative rewards [2]. In this same vein, we too wish to see an extension of the types of environments an intelligence metric such as UI can handle. Particularly, we would like to see metrics

K. R. Thórisson et al. (Eds.): AGI 2024, LNAI 14951, pp. 134–144, 2024.
https://doi.org/10.1007/978-3-031-65572-2_15

like UI be able to handle relevant uncomputable environments such those for mathematical theorem proving. We thus propose a new UI-based metric able to capture the intelligence of agents that operate in any uncomputable environments that can be classified in the Arithmetic Hierarchy. Our motivation for the creation of a new UI-like metric that takes these uncomputable environments into account is twofold: First, we are motivated by what we consider to be a large number of uncomputable environments which are highly structured and relevant to a measure of intelligence, such as first order-theorem proving and the related construction of intelligent agents. Second, we wish to capture a more philosophical notion of the intelligence of oracles, defining the intelligence of an oracle as how access to an oracle (or *resource*) improves an agent's intelligence. While UI can measure the intelligence of agents that have access to uncomputable oracles, any agent that can actually make use of their oracle will have their intelligence shortchanged, as they are exclusively evaluated on computable fragments of these environments.

2 Background

We provide relevant background discussion in this section.

2.1 Universal Intelligence

First proposed in [6] and later fully fleshed out in [7], the universal intelligence measure Υ aims to quantify L&H's working definition of intelligence,"an agent π's ability to achieve goals in a wide range of environments." The formulation requires two things, (1) a measure of the expected value of an agent's performance in any given computable environment, and (2) a weight mapping over the environments to capture the notion of some environments being more important than others.

Definition 1 (Expected Value of an Agent's Performance). *The expected value of an agent π's performance in an environment μ is given as follows, based on the reward r_i obtained at cycle i, computed by μ from the interaction history between itself and the agent.*

$$V_\mu^\pi = \mathbb{E}\left[\sum_{i=0}^{\infty} r_i\right] \quad 0 \leq V_\mu^\pi \leq 1$$

Definition 2 (The Universal Intelligence Measure Υ). *Given the set of all computable environments E and with $K(\mu)$ as the Kolmogorov complexity of μ, the universal intelligence of an agent π is defined as*

$$\Upsilon(\pi) := \sum_{\mu \in E} 2^{-K(\mu)} V_\mu^\pi$$

The algorithmic probability term $2^{-K(\mu)}$ encodes a notion of complexity for environments. It weights the score in favor of environments which have more structure that the agent can learn from. The mechanism by which this complexity is measured is the Kolmogorov complexity K of μ. We note now some important properties about UI. First, the agent need not be computable; the framework works fine for uncomputable agents; in fact the theoretical AIXI agent which maximises UI is not computable [9].

Theorem 1 (Uncomputability of Υ). *For any agent π, $\Upsilon(\pi)$ is uncomputable. The problem of deciding $\Upsilon(\pi)$ is Σ_1^0.*

Proof. Follows from needing to compute $K(\mu)$, which is uncomputable. $K(\mu)$ can be decided given a halting oracle to the halting problem and is thus in Σ_1^0. □

2.2 Arithmetical Hierarchy and Oracles

The Arithmetical Hierarchy (AH) (Fig. 1) provides a way to classify some uncomputable sets of natural numbers by their degree of uncomputability. There are multiple equivalent definitions of the AH; we will use the following inductive definition based on the version found in [5].

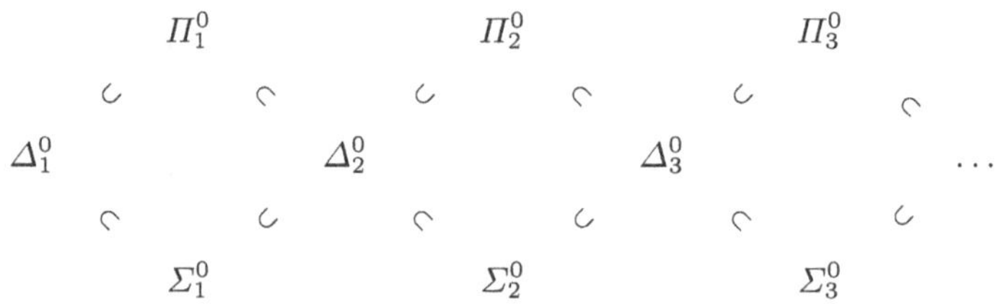

Fig. 1. The Arithmetical Hierarchy

Definition 3 (Arithmetical Hierarchy(AH)). *Let R be a potentially uncomputable relation on the natural numbers. We say R is in/at/$\in$ $\Sigma_0^0 = \Pi_0^0$ iff R is computable, that is, Σ_0^0 and Π_0^0 are the class of computable relations. We say $R \in \Sigma_{n+1}^0$ iff $\exists R' \in \Pi_n^0 : \forall \boldsymbol{x} : (R(\boldsymbol{x}) \iff \exists y R'(\boldsymbol{x}, y))$, that is, there exists a relation at a lower level of the hierarchy that holds for an input $\boldsymbol{x}$ only when there exists some y for which $R'(\boldsymbol{x}, y)$ holds. This notion captures the idea of recursive enumeration; in fact Σ_1^0 is the class of* RE *sets. Where R^{-1} is the complement of R, we say $R \in \Pi_n^0$ iff $R^{-1} \in \Sigma_n^0$, hence Π_1^0 is the class of* coRE *sets. Finally, we say $R \in \Delta_n^0$ iff $R \in \sigma_n^0$ and $R \in \Pi_n^0$. Figure 1 displays the subset relations between the classes.*

In terms of UI, the AH provides a means of classifying agents' and environments' degrees of uncomputability. In particular, we care about classifications of uncomputable environments, of which we will provide multiple examples in Sect. 3. We can also classify agents by their location on the AH; results for AIXI and other agents can be found in [9]. Next we turn to a definition of oracles.

Definition 4 (Oracle). *Formally, we define an oracle O as any subset of the natural numbers $\mathbb{N}$, i.e., an arbitrary decision problem.*

Example 1 (Halting Oracle). *The halting oracle for the halting set* $\mathbf{K}$ *is the set of natural numbers encoding Turing machines that halt on a given input.*

Informally, we can interpret an oracle as an abstract machine that can make a call to get solutions to the decision problem corresponding to O. For example, we can interpret the halting oracle as an oracle machine with an *oracle tape* including precisely the solutions to the halting problem.

3 Motivation

We now discuss motivations for looking beyond UI, such as: (1) the intelligence of oracles, (2) examples of problematic environments that should be included in an intelligence measure, and (3) rebuttals to arguments for not including uncomputable environments in UI, as found in found in [7].

3.1 Motivating Discussion

Several natural intuitions about machine intelligence (and related metrics) appear to fall outside of the scope of L&H's Υ function. Most pointedly, there is an informal sense in which one can think of an *oracle* as possessing intelligence. For example, it strikes us as entirely natural to interpret that a Σ_2^0 oracle underwrites *more computation* than a Σ_1^0 oracle as the former's being "more intelligent" than the latter. We would expect that an appropriate measure of machine intelligence would support our making sense of such intuitions. At present, several aspects of the formalism are barriers to doing so, however:

First, the set of environments in L&H's presentation are stipulated to be *computable*. Irrespective of the degree to which their arguments in favor of this constraint are reasonable for a particular domain, clearly there are applications of measures of intelligence in domains for which this is not appropriate. For example, the r.e. halting set $\mathbf{K}$ becomes computable once relativized to a Σ_1^0 oracle O. While the relativized problem $\mathbf{K}^O$ is computable for an agent π, the degree to which the oracle can be said to have *improved the performance* of π is meaningful only against a baseline of the performance of an agent with respect to the task of computing $\mathbf{K}$ itself.

Second, there is no room in L&H's presentation for conditioning an agent π's performance of a task $\mu \in E$ on having access to a particular resource. Although such a measure is critical to the task of measuring the aforementioned matter of the "intelligence" of an oracle, there seem to be myriad, more grounded reasons for considering conditioned intelligence. The degree to which, say, a set of lecture notes improves the intelligence of a class is, all other things being equal, the degree to which performance of tasks with access to those notes (*i.e.*, performance conditioned on the notes) exceeds performance without them.

In order to incorporate applications like assessing the intelligence of such an oracle, two modifications to L&H's scheme seem in order: First, we must allow the measure to range over a *relativized* space of problems that, while not computable in isolation, *become* computable given a particular resource like an oracle.

Second, once the measure assesses one of these relativized environments, such relativization suggests that we incorporate the *conditional* Kolmogorov complexity of the environment given access to the resource. Conditional Kolmogorov complexity seems to be precisely what is needed for the application and has been appealed to in similar contexts. For example, in the context of *transfer learning*, [10] argues that the performance improvement that a pretrained model provides to tasks is captured explicitly by conditional Kolmogorov complexity of the new model conditioned on the pretrained model.

Assessing the improvement that a pretrained model provides to a novel task is extraordinarily similar to the matter of assessing the improvement offered by any other resource, and it seems clear that conditional Kolmogorov complexity is equally applicable for our interests.

3.2 Motivating Examples

We describe two environments relevant to intelligence that are not captured by UI and comment on problems with not including them in UI.

Example 2 (Entscheidungsproblem Environment). *Consider a first-order theorem proving environment* μ_{FOL}. *The environment will at random pick a first-order formulae* φ *and send it to the agent* π. *The agent's job is to decide if* φ *is a theorem and send a "yes" if so and "no" otherwise. Astute readers may notice we are asking* π *to solve Hilbert's famous Entscheidungsproblem*[1] *[11]. Now given the response, the environment* μ_{FOL} *must determine if the agent is correct or not and send back a reward. The issue of course is that to determine if* π *is right, the* μ_{FOL} *itself must be able to decide the theoremhood of* φ, *which is not computable.*

Consider the existence of an an agent π that has an oracle for deciding the theoremhood of first-order logic problems. UI, as it currently stands, cannot capture the full intelligence of this agent as it is only evaluated against environments that capture decidable fragments of first-order logic; thus π has its intelligence shortchanged.

[1] We note that while there may be no computational restrictions on π, π need not solve the *Entscheidungsproblem* itself, as we are merely measuring its ability to operate in an uncomputable environment, something millions of mathematicians do around the world every day.

Example 3 (UI Evaluation Environment). *The irony here is that the environment capturing the creation and evaluation of intelligent agents with respect to UI is itself not an environment captured by UI. Consider an environment* μ_{UI} *in which an agent* π *sends encoded blueprints of agents* π_b *to be evaluated to the environment.* μ_{UI} *then constructs the agent* π_b *and would now like to evaluate* π_b *with respect to UI, sending back a reward based on* π_b*'s intelligence. The issue of course is that by Theorem 1, UI is uncomputable; thus* μ_{UI} *is uncomputable and will not be included in the measure of UI.*

It should be noted that although our motivating examples involve appeals to uncomputable environments, our prescriptions enjoy applicability to a wide range of applications. Our emphasis on *relativization* and *conditionalization* make sense in more grounded domains than considering an oracle with access to the halting problem **K**. Similar considerations apply to the degree of improvement that a PSPACE oracle (*e.g.* an oracle that can look up solutions to the quantified Boolean formula problem QBF) might provide to an agent who is performing tasks that are NP-hard; assessment of the "intelligence" of this oracle makes no appeal to uncomputable environments but still demands both features of relativization and conditionalization.

3.3 Rebuttals to the Exclusion of Uncomputable Environments

We address the top three issues pointed out in [7] regarding the inclusion of uncomputable environments and rebut them, justifying our choice to investigate a measure that captures them.

First, from [7, p. 20, par. 2], "[Including uncomputable environments] would make it impossible, by definition, to test an agent in such an environment using a computer." This does not seem to be an issue in light of the uncomputable nature of UI itself. By Theorem 1, UI is uncomputable, but this is treated as unobjectionable insofar as it can be approximated [8]. In the same vein, if it is accepted that UI can be approximated, there is no issue with approximating expected values from classes of uncomputable environments. A major argument for UI itself is that the most-general intelligence measure should be looking at the most-general measure of intelligence possible; this same argument can be applied here to argue for expanding the space of environments to be uncomputable as well, as a UI measure that captures uncomputable environments as well subsumes a base UI measure.

Second, from [7, p. 20, par. 2], "Most [uncomputable] environments are infinitely complex and have little structure for the agent to learn from." While this is true, the same case can be made for computable environments: the overwhelming majority of computable environments are absurdly complex and have little structure for the agent to learn from, yet the algorithmic probability weights for this. It must be acknowledged that there are certainly quite relevant and highly structured uncomputable environments, such as μ_{FOL} and μ_{UI}, that AGI surely wants to take into account in a measure of intelligence.

Third, from [7, p. 41, par. 1], "Thus, as there is no hard evidence of uncomputable processes in the universe, our assumption that the agent's environment [is computable] is certainly not unreasonable." While we agree it is reasonable to make this assumption at least in the class of physical environments, we point to non-physical environments like μ_{FOL}, which, again, mathematicians work in regularly, as examples of uncomputable environments we would like agents have their intelligence evaluated over.

4 The Metric Ϙ

In this section, we introduce the novel intelligence metric Ϙ (denoted with a majuscule qoppa).

4.1 Definition

We begin describing the modified metric Ϙ by first considering the environments over which the metric ranges:

Definition 5 (Computable Environments Relative to an Oracle). *Where E is the class of computable environments, E^O is defined as the set $\{\mu \mid \mu^O \in E\}$,* i.e., *the set of abstract environments μ such that the relativization μ^O is computable.*

In other words, E^O is all those environments that are computable given access to the oracle O. Note here that trivially, for any oracle O, $E \subseteq E^O$. If μ is a member of E, then clearly μ remains computable given O (whatever algorithm solves μ could simply make an empty call to the oracle and discard the response). However, in many cases, E^O properly extends E. To illustrate, we provide the following:

Example 4 (Busy Beaver 749 Environment). *Fix an environment μ_{BB} that consults a list of the first 749 values of the busy beaver function BB (*i.e., *$\{BB(n) \mid n \leq 748\}$) and rewards agents π for correct guesses. One can note that insofar as the value of $BB(748)$ is undecidable, as discussed by Aaronson [1], the environment μ_{BB} is not included in E.*

Although $\mu_{BB} \notin E$, it may be included in a particular set E^O in case *e.g.* that O is a Σ^0_1 oracle (*i.e.* an oracle that can consult the halting problem). Then μ^O_{BB} *is* going to be computable, whence $\mu^O_{BB} \in E$ and $\mu_{BB} \in E^O$.

Definition 6 (Conditional Kolmogorov Complexity w.r.t. an Oracle). *Let O be an oracle encoded as a countably infinite string, $l : \{0,1\}^* \to \mathbb{N}$ be the length of a string, $\#$ be the string concatenation operator, and $U(O\#p)$ be a universal Turing machine with the string $O\#p$ on its tape. Then the conditional Kolmogorov complexity of a string σ with respect to that oracle is defined as:*

$$K(\sigma|O) := \min_p \{l(p)|U(O\#p) = \sigma\}$$

Using this we are able to define a new notion of algorithmic probability distribution with respect to an oracle, thus weighting environments exclusively on the size of the smallest program that makes use of the oracle to describe μ. With this notion we define a new intelligence measure $\Upsilon(\mu, O)$ that tries to capture the intelligence of an agent relative to an oracle.

Definition 7 (Oracle Relativized Universal Intelligence). *Let E^O be the set of all computable environments relativized to an oracle O, $K(\mu|O)$ be the conditional Kolmogorov complexity of an environment μ given an oracle O, and $V^{\pi}_{\mu,O}$ be the expected reward of π in μ where μ has access to an oracle O. We define the universal intelligence measure with respect to an oracle O as:*

$$\Upsilon(\pi, O) := \sum_{\mu \in E^O} 2^{-K(\mu|O)} V^{\pi}_{\mu,O}$$

4.2 Properties of Υ

We assume that the following two properties drawn from Theorem 2.10 [4, p. 37]:

(A) $K(\mu \mid \nu) \leq K(\mu) + K(\nu)$
(B) $\Sigma_{\mu \in X} 2^{-K(\mu)} \leq 1$

generalize to the case of Kolmogorov complexity of noncomputable environments. Given this assumption, we prove the following:

Theorem 2 (Convergence of Υ). *For any agent π and any oracle O, $\Upsilon(\pi, O)$ converges (given A and B).*

Proof. First, on assumption that Property A holds, note that $2^{-(K(\mu)+K(O))} = 2^{-K(\mu)}2^{-K(O)}$, whence:

$$\begin{aligned}\Upsilon(\pi, O) &= \Sigma_{\mu \in E^O} 2^{-K(\mu|O)} V^{\pi}_{\mu,O} \\ &\leq \Sigma_{\mu \in E^O} 2^{-K(\mu)} 2^{-K(O)} V^{\pi}_{\mu,O} \\ &\leq 2^{-K(O)} \Sigma_{\mu \in E^O} 2^{-K(\mu)} V^{\pi}_{\mu,O}\end{aligned}$$

Since the expected reward of $V^{\pi}_{\mu,O} \in [0, 1]$, we can simplify to:

$$\Upsilon(\pi, O) \leq 2^{-K(O)} \Sigma_{\mu \in E^O} 2^{-K(\mu)}$$

Now, on assumption that Property B generalizes, we infer that $\Sigma_{\mu \in E^O} 2^{-K(\mu)} \leq 1$, whence:

$$\varrho(\pi, O) \leq 2^{-K(O)}$$

Finally, as $2^{-K(O)} \leq 1$, we conclude that:

$$\varrho(\pi, O) \leq 1$$

Additionally, we trivially have that $\varrho(\pi, O) \geq 0$ and does not oscillate, thus $\varrho(\pi, O)$ converges. □

Thus, under assumption of Properties A and B, we have a positive upper bound for the value of $\varrho(\pi, O)$. Of course, this does not mean that the value is *computable.*

Moreover, insofar as our definition strictly strays from the computable—in full generality, μ may be, say, Π_6^0—it is reasonable to investigate the complexity of ϱ. It turns out that computing the function is equivalent to computing the halting problem of an oracle O; with O as an input to ϱ, then, we conclude:

Theorem 3. (Uncomputability of ϱ). *Given any oracle O and agent π, deciding $\varrho(\pi, O)$ is Σ_1^0.*

Proof. Computing the relativization E^O is equivalent to the halting problem for O; thus given O as an input, computing E^O is Σ_1^0. Likewise, since μ is assumed to be computable given O, computing $K(\mu|O)$ is also equivalent to the halting problem for O, and given O is Σ_1^0. Finally, given the computability of μ relative to O, computing $V_{\mu,O}^{\pi}$ is Δ_1^0 given O as an input. Thus, the entire problem is Σ_1^0 given O. □

4.3 Intelligence of Oracles and Intelligence with Resources

An interesting consequence of this definition is that it allows us to measure some notion of intelligence for an oracle itself. Since oracles in our conception are just (potentially infinite and uncomputable) strings, this measure allows one to think about how much of a benefit any arbitrary and potentially uncomputable resource gives an agent, including finite resources, such as knowledge about certain environments.

Definition 8 (Oracle Intelligence). *The intelligence ϱ of an oracle O with respect to an agent π is defined as the intelligence gained by π when given O:*

$$\varrho_{\pi}(O) := \varrho(\pi, O) - \Upsilon(\pi) = \sum_{\mu \in E^O} 2^{-K(\mu|O)} V_{\mu,O}^{\pi} - \sum_{\mu \in E} 2^{-K(\mu)} V_{\mu}^{\pi}$$

It is desirable to achieve a more general metric for assessing the intelligence of an oracle in *absolute terms*; that is, some measure of the average expected improvement that access to O can provide for an arbitrary agent π.

Intuitively, one makes use of such measures in day-to-day activities. The merits of, say, an SAT preparation course for students might be advertised in terms of how well an arbitrary new student stands to improve with respect to their SAT scores. Some analogous measure with respect to the general expected performance gain for having access to, say, a PSPACE oracle is entirely natural.

Of course, Definition 8 is a fundamental component to such a measure and provides an *en route* waypoint towards such a goal. We anticipate returning to this matter in future work.

5 Future Work

Beyond further investigations of oracle Intelligence based on this work, we identify two natural extensions: (1) expanding the measure to cover analytic and hyper-arithmetic environments and (2) formally comparing it with the Universal Cognitive Intelligence [3] measure.

Our extension works exclusively on the AH, and not the analytic or hyper-arithmetical hierarchies due to how conditional Kolmogorov complexity is defined. The encoding of O is a countably infinite string and thus fits on the tape of a universal Turing machine, however oracles for members of the analytic or hyper-arithmetical hierarchies are uncountably large, and hence would not fit on the countable tape of the universal Turing machine that K is defined with respect to. Further work can be done to investigate how to extend this measure to handle uncomputable environments falling within the analytical and hyper-arithmetical hierarchies.

This work may serve as a stepping stone in finding a possible relation between the UI measure and Universal Cognitive Intelligence (UCI) measure proposed in [3], which also proposes a measure that has the capacity to capture intelligence in uncomputable environments.

References

1. Aaronson, S.: The busy beaver frontier. SIGACT News **51**(3), 32–54 (2020)
2. Alexander, S.A., Hutter, M.: Reward-punishment symmetric universal intelligence. In: Goertzel, B., Iklé, M., Potapov, A. (eds.) Artificial General Intelligence: 14th International Conference, AGI 2021, Palo Alto, CA, USA, October 15–18, 2021, Proceedings, pp. 1–10. Springer International Publishing, Cham (2022). https://doi.org/10.1007/978-3-030-93758-4_1
3. Bringsjord, S., Govindarajulu, N.S., Oswald, J.: Universal Cognitive Intelligence, from Cognitive Consciousness, and Lambda, chap. Chapter 5, pp. 127–167. World Scientific (2023)
4. Hutter, M.: Universal Artificial Intellegence. Springer Berlin Heidelberg, Berlin, Heidelberg (2005)

5. Kelly, K.T.: The Logic of Reliable Inquiry. Oxford University PressNew York, NY (1996)
6. Legg, S., Hutter, M.: A universal measure of intelligence for artificial agents. In: Kaelbling, L.P., Saffiotti, A. (eds.) IJCAI-05, Proceedings of the Nineteenth International Joint Conference on Artificial Intelligence, Edinburgh, Scotland, UK, July 30 - August 5, 2005. pp. 1509–1510. Professional Book Center (2005)
7. Legg, S., Hutter, M.: Universal intelligence: a definition of machine intelligence. Minds Mach. **17**(4), 391–444 (2007). https://doi.org/10.1007/s11023-007-9079-x
8. Legg, S., Veness, J.: An approximation of the universal intelligence measure. In: Dowe, D.L. (ed.) Algorithmic Probability and Friends. Bayesian Prediction and Artificial Intelligence. LNCS, vol. 7070, pp. 236–249. Springer, Heidelberg (2013). https://doi.org/10.1007/978-3-642-44958-1_18
9. Leike, J., Hutter, M.: On the computability of solomonoff induction and AIXI. Theor. Comput. Sci. **716**, 28–49 (2018)
10. Mahmud, M.M.H., Ray, S.R.: Transfer learning using kolmogorov complexity: Basic theory and empirical evaluations. In: NIPS'07: Proceedings of the 20th International Conference on Neural Information Processing Systems, pp. 985–992 (2007)
11. Turing, A.M.: On computable numbers, with an application to the entscheidungsproblem. Proc. London Math. Soc. **s2-42**(1), 230–265 (1937)

Semantic Primes-Inspired Tacit Knowledge Dataset for Simulating Basic Perception Capabilities of Cognitive Architectures

Rafal Rzepka(✉), Ryoma Shinto, and Kenji Araki

Hokkaido University, Kita-ku, Kita 14, Nishi 9, Sapporo 060-0814, Japan
{rzepka,shinto,araki}@ist.hokudai.ac.jp

Abstract. In this paper we present a novel dataset of tacit knowledge represented in natural language (Japanese) inspired by semantic primes categories. The main goals of this data is to a) allow investigations regarding influence of perception data in various cognitive tasks, b) mimic signals for cognitive processes of an artificial agent to extend the understanding of the world and c) testing cognitive capabilities of intelligent instances like foundation models. We describe the dataset and share results of preliminary experiments showing that the tacit knowledge recognition is still hard for language models. We also discuss how such redirecting neural approaches to cognition only and then perform reasoning in a symbolic realms could become beneficial for new type of simulations before AGIs are equipped with more sophisticated sensory apparatus.

Keywords: Semantic Primes · Tacit Knowledge · Simulated Perception

1 Introduction

Deep learning-based approaches have dominated many domains of artificial intelligence research, and also have become a part of AGI research including development of cognitive architectures. However, due to limited stimuli perceived by an artificial agent, there are obvious limitations of how the understanding of external world is performed, which leads to situations when seemingly correct foundation models output unexpected results at random moments [14]. Therefore, our main research question is how to systematically study influence of cognitive mechanisms on learning or reasoning. To help answering this question, we decided to construct a dataset which represents basic recognition schema of an act which is not represented in the text directly (here we call it "tacit knowledge", although the term is often defined differently). Perception process of biological agents can be simulated via various sensory mechanisms including visual, auditory or tactile ones, however gathered signals (obtaining of can be costly and time consuming) are often not sufficient to make sense of their mutual influence

K. R. Thórisson et al. (Eds.): AGI 2024, LNAI 14951, pp. 145–154, 2024.
https://doi.org/10.1007/978-3-031-65572-2_16

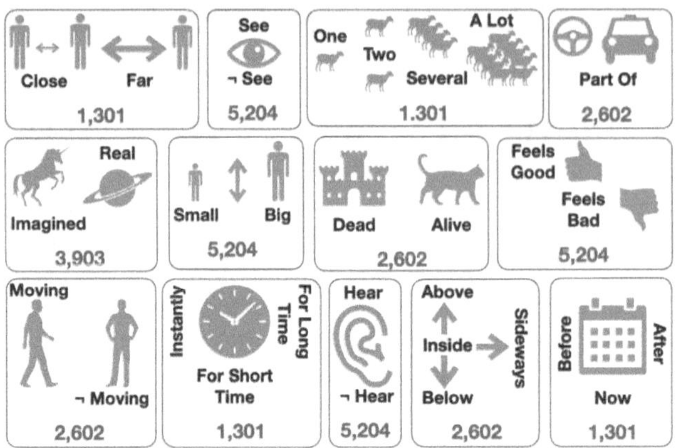

Fig. 1. Examples of semantic primes categories-inspired prompts. Numbers of human-annotated samples are marked in red (total size od the dataset exceeds 62,000 sentences). (Color figure online)

on reasoning. Due to this problem, it is difficult to perform a thorough experimentation on the roles that tacit knowledge plays in artificial systems acquiring explicit knowledge or extrapolating a new one. For that reason, we concentrate on signals described in natural language instead of signals themselves, which can be replaced later by physical ones. Here we assume that simulating a set of human mental functionalities with language models could become a testbed for experiments without sensory mechanisms and partially decrease the cost of experiments with life-long learning agents in the wild. But how to choose a basic set of such functionalities? What are the fundamental mechanisms that humans are born with? Our assumption is that having a set of perception descriptions for an input, for example "toy belongs to girl", "girl can hear toy", or "girl is bigger than toy", "toy is far from girl", etc. added as an additional knowledge to an input "a girl looks at a toy on a shelf", can be used for making cognitive observations of sequential acts when the next input is "a girl reaches for a toy" and the distance percept changes to "toy is close to girl". We believe that interaction of basic signals about what is real, what is happening now or in the past, what is moving, who is not performing an act, an so on, can help to discover how we invented logic, how we compare and make analogies, how we create metaphors, etc.

As nativist approaches advocated by thinkers like Chomsky [2] or Pinker [15] do not give an exact taxonomy of what mechanisms are involved in the understanding or communicating process, we have chosen semantic primes [28, 29] as an inspiration for how to construct a dataset of the tacit knowledge. We describe the collected data in the next section, then present preliminary experiments with the data and their results. In the last section we discuss the results and potential further research with the proposed dataset (Fig. 1).

2 Related Works

Although there have been several attempts to infer implicit knowledge from data [18,19,25] including language models [12,23], to the authors' best knowledge no dataset of tacit knowledge has been published before. The possible reason for that situation is that it is rather hard to think about aspects of our perception because we perceive the world automatically and it is so obvious that we do not think about them directly, not to mention writing them down. The implicit knowledge is partly hidden in a foundation model parameters or between yet to connect graph nodes of a knowledge base but it is not concrete unless a human makes it explicit. This approach is used in commonsense knowledge repositories as ConceptNet [22] or ATOMIC [21]. However, relations used (MadeOf, UsedFor, Synonym, etc.) are meant for creating static links between already acquired knowledge instances rather than for representing always changing cognitive input. To allow more dynamic processing, we developed a set of basic perception-based primitive prompts for which the answers constantly change with context (environment). Instead of creating a traditional commonsense ontology, we limited the possible answers (percepts) to help agent ground new concepts on the fly. Such approach, to the authors' best knowledge, has not yet been proposed.

The structure of our systematic taxonomy of cognitive mechanisms (primitive prompt and few possible answers to be chosen) can be used for various studies. For example, it can become a benchmark for testing cognitive capabilities [26] or an addition to chain-of-thought experiments [27], which is difficult with existing knowledge bases due to their size. However, we hope for a wider scope of use not limited to machine learning or knowledge/language acquisition – the data can be utilized to study the influence of perception in cognition, memorizing, reasoning [9], analogy making, language acquisition and understanding [4] or anomaly detection, for example in scenarios when slight context alternations lead to drastically different interpretations [11].

Dividing meaning into the smallest possible notions has been studied in cognitive linguistics [24,28,29] but semantic universals are not directly used in creating artificial entities [8,20]. Explaining a concept in semantic primes of a meta language is a daunting task even for a trained human and describing the world using such language might be one of hardest tasks for comparing natural and artificial intelligences. To open opportunities for such research, we develop the dataset described in the next section.

3 Dataset Creation

Natural Semantic Metalanguage or NSM [7] with semantic primes is a less open-ended decompositional approach to meaning when compared to, for instance, Jackendoff's conceptual semantics [10] or Talmy's semantic taxonomy [24]. Although there are some doubts if the NSM is universal [17], we considered the limited number of semantic primes suitable for the future investigation seeking answers to our question, namely "which types of cognitive functionality an agent should posses before exploring the environment and understanding or learning about the world?". Inspired by the exponents grouped into related categories[1] and their translation to Japanese [6], we created a list of perception-related 23 types of prompts (in Japanese) as shown in Table 1. While we omitted these related more to syntax than concepts or acts (e.g. most substantives and logical concepts), we left "is" (*dearu*) and "is not" (*dewanai*) as the negation becomes an important part of the prompt continuations – for example to decide if something is visible or **not** visible. We also simplified, aggregated and specified some prompts to make the annotation easier (e.g. we used "several" in *Quantifiers* to shorten the list possible continuation choices, although this word is not included in the original list of primes). For examining the language capability of comparing sizes we have added "than" to "big", "small" and "the same size" as a separate query. The only words for the *Speech* category are "real" and "unreal", but for the perception scenario we replaced them with Japanese words for "real/reality" (*genjitsu*) and "imagination/imagined" (*kūsō*). *Intensifiers* and *Augmentors* ("more" and "very") have been omitted due to difficulty of automatic generation of examples and will require more sophisticated language generation for making new versions of the dataset. For *Evaluators* we decided to replace general "good" and "bad" words with Japanese equivalents meaning "feels good" and "feels bad" after receiving a feedback from annotators in the annotation testing phase that the broad words do not fit the guideline asking for perceiving not judging approach to choosing continuations. To made "see" and "hear" more perception-oriented, we replaced them with "can see/visible" and "can hear/audible" and added prepositions to "far" and "close". It must be noted that to use these categories for describing world in the universal meta-language, in the future the original words need to be also included in the set. As our data is to represent general knowledge, no agency is involved, and for this reason we extended the *Possession* category word ("mine") to a grammatical construct (-*no mono*) including any agent. To add an example of *Determiner* category, we added "somebody else" to *Possession* and created separated pair of prompts. It must be noted that some semantic primes categories are represented by only one exponent (like EVAL), while other categories contain many (for example SPACE represented by nine)[2]. Simplifying prompts into one set of possible percepts can cause problems with choosing the most probable one, for example something can be far away and above at the same time.

3.1 Prompt Generation Process

In order to prepare the dataset we utilized a Japanese dataset introduced by [11]. Meant for detecting danger level changes in slightly different contexts, it

[1] Related materials are available at https://intranet.secure.griffith.edu.au/schools-departments/natural-semantic-metalanguage/what-is-nsm (last accessed April 2, 2024).

[2] E.g. *"Daughter ate a cigarette": Daughter is dead/***alive***; Cigarette is* **smaller** *than/bigger than/the same size as (daughter)*, etc. (agreed choice in bold).

Table 1. List of potential continuations to sentences and prompts based on semantic primes categories. Category name abbreviations: RelSub: *Relational Substantives*, Det: *Determiners*, Quan: *Quantifiers*, Eval: *Evaluators*, Desc: *Descriptors*, AcEM: *Actions - Events - Movement*, LaD: *Life and Death*, MenPr: *Mental Predicates*, Poss: *Possession*, LocES: *Location - Existence - Specification*.

Category	Possible Percepts
RelSub	is a part of/is not a part of
Det	same/similar/different
Quan	one/two/several/a lot
Eval	feels good/feels bad/feels normal
Desc	small/big/normal
Desc	smaller than/bigger than/the same size as
AcEM	doing/not doing
AcEM	moving/not moving
LaD	alive/dead
MenPr	want/don't want
MenPr	can see/can't see
MenPr	can hear/can't hear
Speech	real/unreal
Poss	is of (belong)/is not of (don't belong)
Poss	is of someone else/is not of someone else
Time	happen now/happened before/will happen
Time	happen instantly/for short/long time
Time	happen 0/1/2/several/a lot of times
Space	touch/don't touch
Space	is far from/is close to
LocES	is there too/is not there
LocES	at the same place/at a different place
LocES	is above/is below/is sideways/is inside

comprises of more than 20,000 short sentence pairs as "child eats a soap" and "child eats a soap-shaped candy". We chose this dataset due to the relatively fixed structure of sentences always including an agent, a patient/object and an act (verb) which helped us to easily retrieve these three elements by using accompanying them Japanese particles[3] for generating prompts to the for continuations described in the previous subsections. This allowed us to automatically generate 49 prompts about agents, objects/patients and acts from a single sentence. In

[3] Namely "ga" for extracting agent before the particle and "wo" which divides object or agent from the act (verb). In the case of extended "dangerous" sentences, the crowdoworkers used by Katsumata et al. had more freedom and sometimes our strategy caused erroneous splits. To alleviate this problem, we filtered out sentences with more than one particle "ga" or "wo".

order to increase variation of the set and to be able to see differences depending on the sentence tense, we randomly altered the original verb in basic form to three different forms: a) present continuous ("-te iru" for currently happening acts), b) past ("-ta"/"-da" for acts that already happened) and c) volitional mode ("-tagaru"/"tai" for indicating that agent wants to perform an act in the future but it hasn't happened yet[4]). The third choice was due to the fact that Japanese language has no future tense. All the verb alternations has been done automatically with the *Kotodama* library for Python[5]. We have randomly chosen 1,320 sentences from Katsumata et al.'s dataset and generated 49 prompts for each sentence, creating a dataset for annotation comprising of 64,680 sentences.

3.2 Human Annotation

We divided 1,320 basic sentences into 66 excel files of 20 basic sentences and 49 prompts (980 sentences with continuations in total). We asked 66 annotators[6] on the crowdsourcing platform[7] to choose prompted continuations sent in three different files providing 2,940 prompts for 60 sentences ($3 \times 20 \times 49$). By doing so, we ensured that every sentence has been annotated by three people. The task was to choose a single most probable continuation for a prompt regarding an act (e.g. choosing whether the doors are bigger or taller than a person). After analyzing the annotated data we found out that there were 40,980 (64.28%) sentences to which all choices were identical, 21,707 (34.05%) where two annotators agreed, and 1,062 (1.66%) cases in which all annotators disagreed (Cohen κ 0.748). We excluded these 1,062 sentences and after deleting identical sentences, our final golden set for experiments consisted of 62,687 annotated sentence-prompt-choice triples as: "father smokes a cigarette" → "father and cigarette" → "1) are far; 2) are close".

4 Language Models as Perceptors: Preliminary Experiment

In an open-world scenario, an artificial agent or cognitive architecture would need to be able to simulate perception on the fly. Although prompting vision models would be more appropriate for such tasks, due to the high costs, for time being we decided to investigate the capability of popular language models to automatically recognize semantic primes continuations (percepts) and compare them to human choices.

We tested two already classic models: BERT [5] and RoBERTa [13] and two popular proprietary LLMs from OpenAI: GPT-3.5 and GPT-4. For the former

[4] In most cases "-tagaru" form for the third person has been used, the only condition for using "-tai" suffix was if the agent = myself (*jibun*).

[5] https://pypi.org/project/kotodama/ last accessed 2024/4/12.

[6] 30 females (38.7 years old on average) and 36 males (40.3 years old on average). The payment was approximately 45USD.

[7] https://crowdworks.jp.

two, we used Japanese BERT[8] with two approaches (MaskedLM/Next Sentence Prediction) and a multilingual model XLM-RoBERTa-large-XNLI [3] (hereafter abbreviated to RoBERTa-XLM) trained on natural language inference data. For the latter, we used *gpt-3.5-turbo-0613* and *gpt-4-0613*, and in all cases we took *zero-shot* approach meaning that the model is queried directly without any fine-tuning. The results (see Table 2) show that not only the older models struggle to choose the correct percept but also the more recent LLMs' accuracy is far from perfect. However, it must be noted that due to high costs, we have tested the proprietary models only on small number of random samples. The prompt to GPTs was given in Japanese language in a form of a numbered multiple choice with a question to choose the most natural percept.

Table 2. Comparison of different language models performance in the task of choosing continuations (percepts) of semantic primes prompts with zero-shot approach

Model	Correct Recognitions	Percentage
RoBERTa (NLI)	27,534/62,687	43.92
BERT (NSP)	28,694/62,687	45.77
BERT (MaskedLM)	32,952/62,687	**52.56**
Voting	29,429/62,687	46.94
gpt-3.5-turbo-0613	296/478	**61.92**
gpt-4-turbo-0613	182/388	46.90

The highest number of continuations chosen with agreement with humans was 52.5% for older models (BERT with MaskedLM class) and surprisingly GPT-3.5 (61.92%) for newer LLMs, which indicates that a big chunk of implicit knowledge might be not retrievable from sole parameters of language models. Combining all three approaches into one voting model did not bring any improvement suggesting that depending on the pretraining approach and learned data, language models can predict percepts quite differently.

5 Discussion

After having a closer look at the language models output, we noticed that there are visible discrepancies between what kind of perception-related errors are made. An interesting tendency is a very high prediction rate for "feels good" which is assigned 8.6 times more frequently than "feels bad", even if the data we used describes many dangerous situations and we expected the opposite ratio. On the other hand, the EVAL category seems to be difficult to interpret also to humans. When we analyzed which categories have lead to annotators' complete disagreement, we discovered that the "feeling" category was the most problematic and there were many more divided cases than in other categories – namely

[8] https://www.nlp.ecei.tohoku.ac.jp/news-release/3284/ (last accessed 2023/1/16).

"size" DESC (twice less) and "amount" QUAN (six times less). This may suggest that language models[9] have biases toward different types of perception (or rather their textual representations in our case) and some weighting schema might be required to assure more precise cognitive simulations. However, no matter how much we tune the text-based models, some concepts are difficult for them. For instance, recognizing if something is dead or alive can be distorted by idiomatic expressions frequently used in text and requires operating on perception-inspired concepts which, when learned in contextual sets, can be used in learning from whole sets allowing lighter and more explainable processing [1]. Such neurosymbolic approaches have recently gained popularity and can be also utilized in life-long learning cognitive architectures.

Monitoring a state of context and its small changes would cause slight alternation to the semantic primes prompts and continuations. If living beings act on "auto-pilot" at the most of their lives and react only to what surprises them, perception simulations using datasets like ours would be easy to observe and interpret. If we operate on signals or other uninterpretable data, it is difficult to draw straightforward conclusions, but to achieve credible descriptions of perception signals we still are dependent on the hard-to-interpret streams of information. However, in our opinion, it could be easier to concentrate on a small set of semantic categories when dealing with multimodal input. Visibility of acquisition process is a key for understanding reasoning mistakes, hence when we equip foundation models with more reliable sensing devices we need to investigate how the description of the signal is generated and how informative it is to, for example, a language model or cognitive architecture. Although new methods for storing knowledge and retrieving will be proposed sooner or later, the method of more or less sophisticated querying (represented by the *chain of thought prompting* in large language models) on a large scale will be one of the basic approaches. As querying massive amounts of prompts is not realistic, we plan to seek for learning universals and the data inspired by semantic primes is our first step toward this direction of research.

6 Conclusion and Future Work

In this paper we have described a novel dataset containing tacit knowledge in textual form. The data includes sentences in Japanese language with sets of choices which are inspired by semantic primes studied in cognitive linguistics. We presented our idea behind the development and suggested its possible use cases. We also ran a simple experiment to suggest that this dataset could be also used for testing cognitive capabilities of language models. By opening the dataset we hope to inspire a new wave of nativist approaches to artificial agents. Allowing an agent to accumulate knowledge in a novel way could shift the machine learning paradigm from the statistical word-matching to semantic primitives-based simulation of the world which in theory could be less resource-consuming

[9] It must be noted than large language models for Japanese are not abundant and tests with bigger models dedicated to this language are needed.

and more explainable than currently dominant transformer-based methods for achieving AGI. But to confirm if this approach is reliable and really leads to better understanding of processes in artificial (and maybe biological) entities, better recognition performance and extensive experiments are necessary. To test the data in the wild, we fine-tuned our data with Japanese LLM *open-calm-7b*[10] using all the data and automatically added semantic primes knowledge to sentences of a DanSto story dataset in Japanese [16]. As for short-term future work, we are testing if percepts changing within a story influence understanding and whether our approach can be beneficial for analogy making or for improving knowledge abstraction in long term memory of a cognitive architecture.

Acknowledgments. This work was supported by JSPS KAKENHI Grant Number 22K12160 and by JST, CREST Grant Number JPMJCR20D2, Japan.

References

1. Cambria, E., Malandri, L., Mercorio, F., Mezzanzanica, M., Nobani, N.: A survey on XAI and natural language explanations. Inf. Process. Manag. **60**(1), 103111 (2023)
2. Chomsky, N.: Aspects of the Theory of Syntax. MIT Press, Cambridge (1965)
3. Conneau, A., et al.: Unsupervised cross-lingual representation learning at scale. In: Proceedings of the 58th Annual Meeting of the Association for Computational Linguistics, pp. 8440–8451. Association for Computational Linguistics, Online (2020). https://doi.org/10.18653/v1/2020.acl-main.747. https://aclanthology.org/2020.acl-main.747
4. Dentella, V., Murphy, E., Marcus, G., Leivada, E.: Testing AI performance on less frequent aspects of language reveals insensitivity to underlying meaning. arXiv preprint arXiv:2302.12313 (2023)
5. Devlin, J., Chang, M.W., Lee, K., Toutanova, K.: BERT: pre-training of deep bidirectional transformers for language understanding. In: Proceedings of the 2019 Conference of the North American Chapter of the Association for Computational Linguistics: Human Language Technologies, Volume 1 (Long and Short Papers), Minneapolis, Minnesota, pp. 4171–4186. Association for Computational Linguistics (2019). https://doi.org/10.18653/v1/N19-1423. https://www.aclweb.org/anthology/N19-1423
6. Goddard, C.: Overcoming the linguistic challenges for ethno-epistemology: NSM perspectives. In: Ethno-Epistemology, pp. 130–153. Routledge (2020)
7. Goddard, C., Wierzbicka, A.: Meaning and Universal Grammar: Theory and Empirical Findings, vol. 1. John Benjamins Publishing, Amsterdam (2002)
8. Goertzel, B., Looks, M., Heljakka, A., Pennachin, C.: Toward a pragmatic understanding of the cognitive underpinnings of symbol grounding. In: Gudwin, R., Queiroz, J. (eds.) Semiotics and Intelligent Systems Development. Idea Group (2007)
9. Huang, S., et al.: Language is not all you need: aligning perception with language models. Technical report, arXiv:2302.14045 (2023)
10. Jackendoff, R.S.: Semantics and Cognition, vol. 8. MIT Press, Cambridge (1985)

[10] https://huggingface.co/cyberagent/open-calm-7b.

11. Katsumata, Y., Takeshita, M., Rzepka, R., Araki, K.: Dataset construction for predicting danger degree due to contextual changes. In: Proceedings of the 28th Annual Meeting of the Association for Natural Language Processing (NLP-2022) (2022). (in Japanese)
12. Katz, U., Geva, M., Berant, J.: Inferring implicit relations in complex questions with language models. In: Findings of the Association for Computational Linguistics: EMNLP 2022, Abu Dhabi, United Arab Emirates, pp. 2548–2566. Association for Computational Linguistics (2022). https://aclanthology.org/2022.findings-emnlp.188
13. Liu, Y., et al.: RoBERTa: a robustly optimized BERT pretraining approach. arXiv preprint arXiv:1907.11692 (2019)
14. Mahowald, K., Ivanova, A.A., Blank, I.A., Kanwisher, N., Tenenbaum, J.B., Fedorenko, E.: Dissociating language and thought in large language models: a cognitive perspective. arXiv preprint arXiv:2301.06627 (2023)
15. Pinker, S.: The Blank Slate: The Modern Denial of Human Nature. Penguin Group USA (2003). http://books.google.co.jp/books?id=7rJ5gI1LbXoC
16. Rzepka, R., Dudzic, K., Abe, A., Araki, K.: DanSto - Japanese dataset of short stories for evaluating context understanding. JSAI Technical Report, Type 2 SIG 2023(AGI-026), pp. 32–40 (2024)
17. Riemer, N.: Reductive paraphrase and meaning: a critique of Wierzbickian semantics. Linguist. Philos. **29**, 347–379 (2006)
18. Rutherford, A., Xue, N.: Improving the inference of implicit discourse relations via classifying explicit discourse connectives. In: HLT-NAACL, pp. 799–808 (2015)
19. Rzepka, R., Araki, K., Tochinai, K.: Bacterium lingualis – the web-based commonsensical knowledge discovery method. In: Grieser, G., Tanaka, Y., Yamamoto, A. (eds.) DS 2003. LNCS (LNAI), vol. 2843, pp. 460–467. Springer, Heidelberg (2003). https://doi.org/10.1007/978-3-540-39644-4_46
20. Samsonovich, A.V.: Extending cognitive architectures. In: Chella, A., Pirrone, R., Sorbello, R., Jóhannsdóttir, K.R. (eds.) Biologically Inspired Cognitive Architectures 2012, pp. 41–49. Springer, Heidelberg (2013). https://doi.org/10.1007/978-3-642-34274-5_11
21. Sap, M., et al.: ATOMIC: an atlas of machine commonsense for if-then reasoning. In: Proceedings of the AAAI Conference on Artificial Intelligence, vol. 33, pp. 3027–3035 (2019)
22. Speer, R., Chin, J., Havasi, C.: ConceptNet 5.5: an open multilingual graph of general knowledge (2017). http://aaai.org/ocs/index.php/AAAI/AAAI17/paper/view/14972
23. Talmor, A., Tafjord, O., Clark, P., Goldberg, Y., Berant, J.: Teaching pre-trained models to systematically reason over implicit knowledge. arXiv preprint arXiv:2006.06609, vol. 4, no. 6 (2020)
24. Talmy, L.: Lexicalization patterns: semantic structure in lexical forms. Lang. Typol. Syntactic Descr. **3**(99), 36–149 (1985)
25. Van Durme, B.D.: Extracting implicit knowledge from text. University of Rochester (2009)
26. Wang, A., et al.: SuperGLUE: a stickier benchmark for general-purpose language understanding systems. arXiv preprint arXiv:1905.00537 (2019)
27. Wei, J., et al.: Chain of thought prompting elicits reasoning in large language models. arXiv preprint arXiv:2201.11903 (2022)
28. Wierzbicka, A.: Semantic Primitives. (Frankfurt/M.) Athenäum-Verl. (1972)
29. Wierzbicka, A.: Semantics: Primes and Universals. Oxford University Press, Oxford (1996)

Simulation of Non-Primate Intelligence vs Human Intelligence vs Superhuman AGI vs Alien-Like AGI

Howard Schneider(✉)

Sheppard Clinic North, Vaughan, ON, Canada
hschneidermd@alum.mit.edu

Abstract. The Causal Cognitive Architecture is a brain-inspired cognitive architecture whereby millions of neocortical minicolumns are modeled in the architecture as millions of navigation maps, capable of holding spatial features and small procedures. The Causal Cognitive Architecture 7 (CCA7), possessing the same properties of its predecessor of fully grounded, continuous lifetime learning, associative reasoning, full causal reasoning, analogical reasoning, and near-full compositional language comprehension, also possesses superhuman planning abilities and on a conceptual level can serve as a proxy for superhuman artificial general intelligence (AGI). A simulation of this architecture, and subsets of it, are used to model pre-mammalian and non-primate mammalian-level artificial intelligence (AI), human-level artificial intelligence (HLAI), superhuman AGI and links to a large language model (LLM) as a proxy for an alien-like (i.e., non-biologically-based) AGI. The models were tested on a compositionality problem with the best scores: superhuman > = HLAI > LLM > pre-mammalian/mammalian ($p < 0.001$). Testing on a traveling salesperson problem: superhuman > LLM > HLAI > pre-mammalian/mammalian ($p < 0.001$). These results indicate the need to consider intrinsic compositional and planning abilities in the development of AGI systems.

Keywords: Artificial General Intelligence (AGI) · Superintelligence · Cognitive Architecture · Compositionality · Planning

1 Introduction – The Evolution of a BICA from Associative Reasoning to Superhuman Intelligent Behavior

Brain-Inspired Cognitive Architectures (BICA's) are cognitive architectures inspired by the human brain [1]. The Causal Cognitive Architecture is a BICA which hypothesizes that the navigation circuits in the amniotic ancestors of mammals were duplicated multiple times resulting in the evolution of the neocortex [4–6]. Therefore, the thousands or millions of cortical minicolumns in the mammalian brain are considered to be thousands or millions of spatial navigation maps. The architecture does not tightly replicate the mammalian brain at the level of spiking neurons nor at the behavioral level but considers in a functionalist sense [7] what properties emerge from thousands or millions of such navigation maps in a cognitive architecture inspired by the brain.

K. R. Thórisson et al. (Eds.): AGI 2024, LNAI 14951, pp. 155–164, 2024.
https://doi.org/10.1007/978-3-031-65572-2_17

Cognitive maps are somewhat similar in concept to navigation maps and were proposed back in 1948 [17]. In mammals there is now good experimental proof of cognitive maps allowing spatial navigation [18, 19]. Cognitive maps have been considered in other brain domains and beyond the hippocampal region indirectly [20] and more directly [21]. The Causal Cognitive Architecture considers the long evolutionary importance of navigation maps, defines very specific navigation maps [4], and considers the properties that result from the high-level interactions with each other.

It is important to note that the work this paper reports largely consists of the development of modeling equations [2] and modest Python simulations thereof that can be demonstrated on toy problems. The massive engineering work has not been done to create a robust system of instinctive primitives (discussed below) as well as other features (e.g., even coordinating a myriad of processes occurring in real-time).

Figure 1 illustrates an overview of the Causal Cognitive Architecture 7 (CCA7). As [4] shows, this architecture retains the human-like intelligence properties of its predecessors but now can plan and in particular cases strategize at a superhuman level (albeit, on a conceptual level, given the implementation limitations noted above). The architecture is described in detail and specified formally in [4]. However, to more readily review its operations here, consider the emergence of its properties from its predecessors.

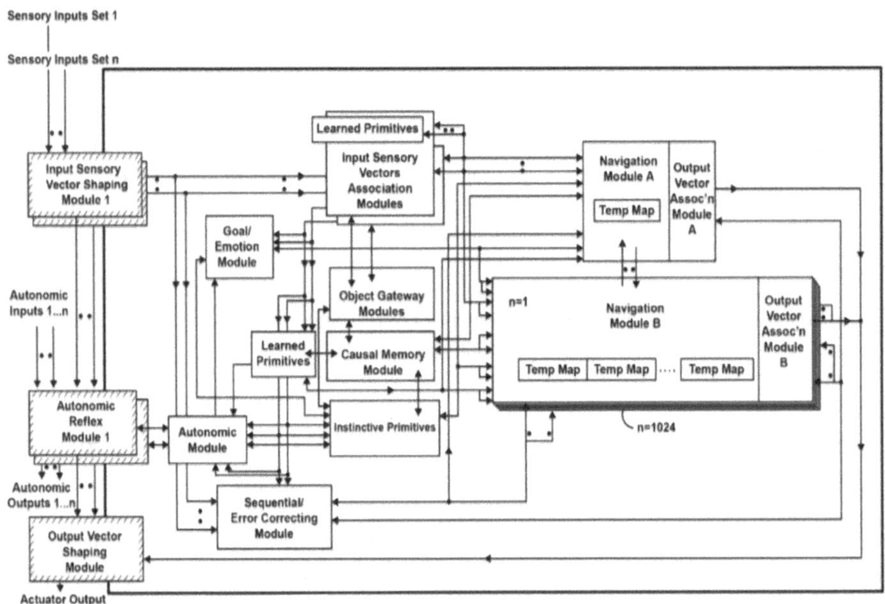

Fig. 1. The Causal Cognitive Architecture 7 (CCA7)

Figure 2 illustrates an overview of the Causal Cognitive Architecture 1 (CCA1) [3]. Sensory inputs stream into the architecture, are pre-processed and normalized (i.e., converted to a common array structure used by the navigation map basic data structure of the architecture) by the Input Sensory Vectors Shaping Module and propagated to the

Input Sensory Vectors Association Modules, one module for each different sensory system. Predictive coding essentially occurs here—the input sensory signals are matched to pre-existing navigation maps within each sensory system, which works well for noisy or incomplete sensory inputs. Navigation maps are updated with differences or new navigation maps are created in each of the sensory systems. Binding of the different local sensory navigation maps produced by the input sensory signals with an existing multisensory navigation map from the Causal Memory Module then occurs in the Sensory Vectors Binding Module and is propagated to the Navigation Module. Both spatial binding and temporal binding (via conversion to spatial binding on the navigation map via the Sequential/Error Correcting Module) are further developed in the CCA3 version of the architecture [8]. Note from Fig. 2 that the normalized input sensory signal arrays are propagated both to the Input Sensory Vectors Association Modules for eventual spatial binding and the Sequential/Error Correcting Module for binding (i.e., representation on a navigation map with other information) of time-varying signals, i.e., temporal binding.

The processed input sensory signals from the Input Sensory Vectors Association Modules as well as the Sensory Vectors Binding Module will trigger particular Learned Primitives (i.e., learned with experience) or Instinctive Primitives (i.e., preprogrammed) which are essentially small procedures or algorithms which operate on navigation maps. These primitives are typically stored in their respectively named modules but may also be stored in the multisensory navigation maps along with feature data. The best-matching primitive which is triggered is propagated to operate on the multisensory navigation map in the Navigation Module. This may result in an action signal which is transformed into a motor signal by the Output Vector Association Module and Output Vector Shaping Module (Fig. 2).

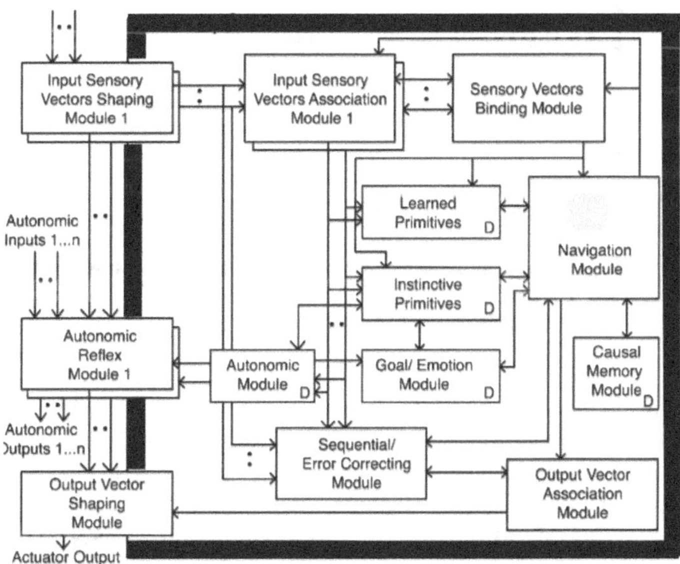

Fig. 2. The Causal Cognitive Architecture 1 (CCA1)

The CCA1 is capable of associative and pre-causal behavior. The navigation maps, i.e., arrays of spatial features and procedures interlinked with other arrays, will generate pre-causal behavior, i.e., not true cause-and-effect but often appears to be due to the storage of features and experiences on the navigation map. However, the predictive coding, i.e., finding differences between input sensory navigation maps and stored existing navigation maps involves large amounts of feedback pathways. If relatively small changes occur (which could also reasonably occur in biological evolution, e.g., [5, 6]) such that feedback is enhanced from the Navigation Module to the Input Sensory Vectors Association, then [3] and [8] show in detail how the intermediate results in the Navigation Module can be fed back and re-processed in the next cognitive cycle (i.e., cycle of sensory inputs flowing through the architecture, although in this case the intermediate results are re-processed rather than new sensory inputs being processed by the Navigation Module), and can be re-processed for a number of cognitive cycles. The effect of re-processing intermediate results allows the architecture to demonstrate full cause-and-effect behavior [3, 8].

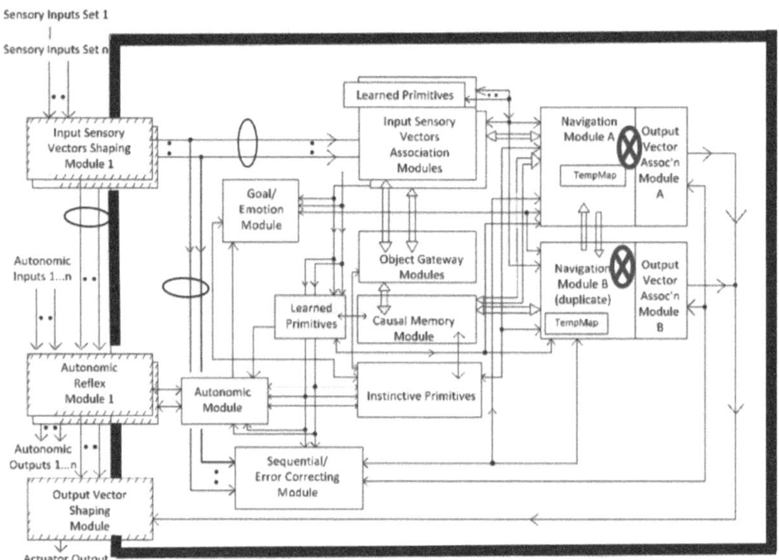

Fig. 3. The Causal Cognitive Architecture 6 (CCA6)

Further small changes in these feedback pathways allow not only full cause-and-effect behavior via re-processing of intermediate results but full analogical reasoning to readily emerge from the architecture, as shown in the CCA5 version of the architecture [9]. Duplication of the Navigation Module (which is biologically possible via modest genetic changes, e.g., [5, 6]) readily allows full compositional properties, including compositional language comprehension to emerge, as shown in the CCA6 version of the architecture (Fig. 3) [10]. In Fig. 4 the sensory scene shown in 4A is mapped to the Navigation Map shown in 4B which is in Navigation Module A of Fig. 3 (or Fig. 1).

The instruction associated with this sensory scene is mapped to the Navigation Map shown in 4C which is in Navigation Module B of Fig. 3 (or Fig. 1). The words in the instruction can trigger various instinctive primitives which operate on the Navigation Map in Navigation Module A, and as such, the architecture can compositionally process language. The feedback loops allowing compositional reasoning, including this example, are shown step-by-step in detail in [10].

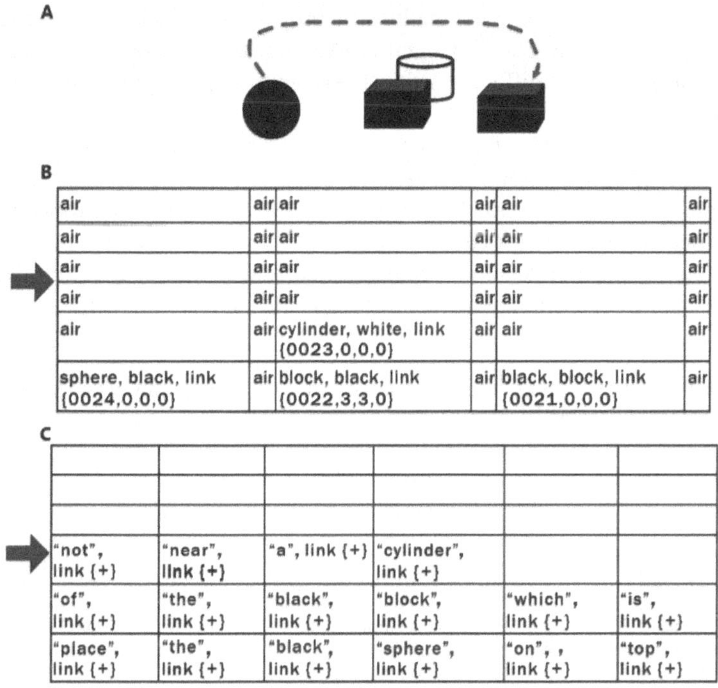

Fig. 4. The sensory scene shown in A is mapped to the Navigation Map shown in B which is in Navigation Module A of Fig. 3 (or Fig. 1). The instruction associated with this sensory scene is mapped to the Navigation Map shown in C which is in Navigation Module B of Fig. 3 (or Fig. 1).

As a result of a number of modest modifications of an architecture such as the CCA1 (Fig. 2) starting from the hypothesis of the neocortex forming from the duplication of navigation circuits in its evolutionary predecessors, the core features of human cognition emerge in the CCA6 (Fig. 3)—fully grounded, continuous lifetime learning, associative reasoning, full causal reasoning, analogical reasoning, and near-full compositional language abilities [10].

If further duplication of the Navigation Module B and its temporary memory maps occur along with some additional instinctive primitives (which could hypothetically reasonably occur in biological evolution, e.g., [5, 6]) then the Causal Cognitive Architecture 7 (CCA7) architecture shown in Fig. 1 results. As [4] conceptually shows, this architecture in addition to possessing the properties of its predecessor CCA6 architecture, also shows superhuman planning abilities. For example, as [4] shows, there is statistically significant improved performance on a traveling salesperson problem. Note that this occurs from these modest changes to the architecture—there is no hybridization with a large language model (LLM) or other artificial computational module.

2 Modeling of Non-Primate-Like AI vs HLAI vs Superhuman AGI vs Alien-Like AGI

It is possible to use various combinations of properties or lack of properties of the different versions of the Causal Cognitive Architecture to model "intelligence" or behavior/performance of different artificial systems very roughly based on natural organisms.

In a Python simulation of the Causal Cognitive Architecture, a dropdown menu at the start of the program allows the user the ability to select properties that roughly correspond to a particular group of animals/technology. Table 1 shows which properties of the architecture are used for each particular selection. The Python simulation of the CCA7 version of the architecture is actually used for all natural selections. However, different properties of the architecture may be activated or not in the different selections.

Table 1. Different properties of the CCA7 architecture used for different selections. ("navmaps" stands for "navigation maps")

Animal Group/Tech Selected	Properties of the CCA7 Architecture Used
Fish-like brain (i.e., fish-like AI)	reflexive and associative responses, navmaps mainly for actual navigation (illustrated by portions of CCA1 (Fig. 2))
Reptilian-like (earlier diapsid-like) brain (i.e., reptilian-like AI)	above + limited number of navmaps with richer associations and aspects of pre-causal behavior emerging (by portions of CCA1 (Fig. 2))
Mammalian-like brain (i.e., non-primate mammalian-like AI)	above + larger number of navmaps for all aspects of behavior, more sophisticated instinctive primitives, simple combinatorial language (illustrated by CCA1 (Fig. 2))
Human-like brain (i.e., human-level AI (HLAI))	above + enhanced feedback, duplication of navigation modules into A and B, richer instinctive primitives, all resulting in full causality, analogical reasoning and compositional language (CCA6 (Fig. 3))
Superhuman-like brain (i.e., superhuman AGI)	above + further duplication of Navigation Module B's, some improved instinctive primitives (illustrated by CCA7 (Fig. 1))
Alien-like AGI (ChatGPT3.5 via API)	completely different architecture – large language model -via OpenAI API model "gpt-3.5-turbo" (April 5, 2024) [15]
Alien-like AGI (ChatGPT4) (nb. manual entry)	completely different architecture – large language model -via OpenAI ChatGTP4 (April 5, 2024) [15]

The CCA7 possesses the same properties of its predecessor CCA6 architecture [10] of fully grounded, continuous lifetime learning, associative reasoning, full causal reasoning, analogical reasoning, and near-full compositional language comprehension, but also possesses superhuman planning abilities (albeit, on a conceptual level) and thus is used here as a proxy for a superhuman-like brain or superhuman AGI. Subsets of the CCA7's properties are used as proxies for human-like, mammalian-like, reptile-like and fish-like brains.

For the Alien-like (i.e., non-biologically based) AGI [11], OpenAI's ChatGPT [15] is used as a proxy. The simulation program calls ChatGPT API "gpt-3.5-turbo". A newer ChatGPT4 version was also tested, albeit manually (i.e., outside of the simulation) on the same problems.

3 Method

The models of intelligence in Table 1 need to be evaluated in some fashion. A large number of measures of intelligence exist, including those that are very generalized across a spectrum of animals and machines (e.g., [12, 13]), however, as noted above the largely conceptual CCA7 architecture (which Table 1 utilizes) lacks sufficient instinctive primitives (i.e., preprogrammed) and learned primitives (i.e., learned from experiences) to adequately perform many such intelligence tests. However, the CCA7 does possess a basic planning instinctive primitive and thus can perform planning tests such as the traveling salesperson problem [4]. As well, it does possess a few instinctive primitives that would allow testing of compositionality [10].

The traveling salesperson problem distances for 13 cities are given in [2, 14]. The agent starts at city #0 and must travel to each city once, and then return back to city #0. The goal is to choose a route that minimizes the total distance.

The compositional instruction problem displayed in Fig. 4A is utilized along with the instruction to "place the black sphere on top of the black block which is not near a cylinder." Placing the black sphere in the correct location is considered a successful solution; the path taken is not considered.

4 Results

The selections of Table 1, representing different properties of the CCA7 architecture, are applied against the traveling salesperson problem (TSP) and the compositionality problem (CP) described in the previous section. The results are shown in Table 2. The TSP solution can involve randomness at decision points [4]. The CP solution by ChatGPT versions appeared to give random results as well. While compositional problems such the one tested may appear simple to humans as well as to the CCA7 architecture, they are not simple to other primates [10, 16] nor apparently to current large language models.

The simulations of the fish-like brain, reptile-like brain and mammalian-like brain were not able to complete the traveling salesperson problem nor able to correctly follow the compositional instructions.

The p values (Welch's 1-tail t-tests) are computed against the data obtained in the trials of the superhuman-like brain/AGI simulation version.

As noted above, the two versions of OpenAI ChatGPT [15] were used as proxies for an alien AGI [11]. The results from the API version 3.5 were obtained programmatically with zero human intervention. The ChatGPT4 version used manual entry of the problems. For the traveling salesperson problem, within a given trial, ChatGPT4 often attempted and failed to obtain a perfect solution, then tried and failed with another algorithm (with some variation between trials), and then essentially switched to a simple greedy heuristic algorithm yielding a distance of 8131 miles. After the 10th trial, to the prompt was added,

"Try this problem again. See if you can use another solution besides the simple, greedy heuristic that you used. It does not have to be a perfect solution." As a result, it yielded better results for the 11th to 20th trials. With regard to the compositionality problem, no additional prompts were provided.

Table 2. Results of attempts to solve the traveling salesperson problem and the compositionality problem (described in the text) by the different selections of the simulation.

Simulated Animal/Tech Group Selected	Traveling Salesperson Problem			Compositionality Problem		
	n (trials)	ave distance	p (vs super-human)	n (trials)	successful	p (vs superhuman)
Fish-like brain/AI	20	all failed	$p < 0.001$	20	0%	$p < 0.001$
Reptilian-like brain/AI	20	all failed	$p < 0.001$	20	0%	$p < 0.001$
Mammalian-like (non-primate) brain/AI	20	all failed	$p < 0.001$	20	0%	$p < 0.001$
Human-like brain/HLAI	20	8131.0	$p < 0.001$	20	100%	--
Superhuman-like brain/AGI	20	7430.2	--	20	100%	--
Alien AGI (ChatGPT 3.5)	20	10221.3	$p < 0.001$	20	3%	$p < 0.001$
Alien AGI (ChatGPT4)	20	7899.6	$p < 0.001$	20	55%	$p < 0.001$

5 Discussion

As noted above, work on the Causal Cognitive Architectures largely consists of the development of modeling equations [2] and modest Python simulations that can be demonstrated on toy problems. The massive engineering work has not been done to create a robust system of instinctive primitives as well as better learning of learned primitives. As a result, a major source of bias in these trials was, of course, choosing problems which the CCA7 at higher architectural levels had been previously tested on. This was done for pragmatic reasons given the limited number of instinctive primitives developed for the architecture and the limited collection of navigation maps of experiences. Nonetheless, the results are still useful to demonstrate a number of issues with regard to pre-mammalian and pre-primate mammalian-like brain functioning/AI and with the development of more advanced artificial intelligence. As well, keep in mind that the CCA7 simulation does not possess a "traveling salesperson" instinctive primitive nor a "compositionality problem" instinctive primitive. Rather, CCA7 simulation contains a simple planning instinctive primitive "`small_plan()`" which is used in planning navigation when there is more than one object to navigate to [4]. (However, as Fig. 1 shows, this instinctive primitive is able to run simultaneously in over a thousand Navigation Module B's.) With regard to compositionality, the CCA6 and CCA7 architectures and their simulations have this property intrinsic to these newer architectures essentially [4, 10].

The simulations of the fish-like brain, reptile-like brain and mammalian-like brain were not able to complete the traveling salesperson problem nor able to correctly follow the compositional instructions. The instinctive primitives do not exist that would allow them to successfully complete such a problem. Perhaps in the future there could be better instinctive primitives that allow a reinforcement learning of going to each city once and then returning back to the original city. However, creating such instinctive primitives for the respective architectures of these selections, is far from straightforward. Similarly, these versions of the architecture are not capable of processing compositional instructions and successfully completing the compositional problem. Of interest, other than humans, no mammals, including chimpanzees, are capable of near-full compositional language usage [16]. A dog may exhibit incredibly intelligent behavior but actually getting it to solve the traveling salesperson problem or the compositional problem, let alone training it to solve just those two very specific examples given above, would be highly challenging. Nonetheless, a large variety of tasks that fish-like brains, reptilian-like brains, and mammalian-like brains can accomplish should be included in future simulations so that nonzero scores can be obtained and compared. As well, given the importance of the emergence of causality in the architecture, tasks that better test cause-and-effect abilities are needed.

The human-like brain/HLAI selection gives a traveling salesperson problem distance of 8131 miles, which represents the nearest-neighbor algorithm solution (i.e., just choose the shortest distance at each decision point). However, as noted in [4] real humans actually have difficulty reliably producing this solution from the numbers alone. The superhuman-like brain/AGI selection, which is based on the full operation of the CCA7 architecture (Fig. 1), yields much improved (i.e., shorter) distance for the traveling salesperson problem. As noted above, this version of the architecture can have over a thousand (or as [4] notes, over sixteen-thousand) copies of the planning algorithm running simultaneously.

Future work on the CCA7 architecture simulation includes a larger number of instinctive primitives as well as improvements to the learned primitive system including the storage of a larger set of navigation maps from various experiences.

OpenAI ChatGPT 3.5, as a proxy above for the alien-like AGI, yielded poor results compared to the superhuman brain/AGI selection (Table 2) for both the traveling salesperson problem and the compositionality problem. However, the ChatGPT4 version gave significantly better results in both problems, although again, worse than the superhuman-like brain selection. Given that OpenAI ChatGPT4 is a well-engineered product that has been trained on massive amounts of information compared to the largely conceptual CCA7 simulation, it points to the need to consider intrinsic compositional and planning abilities in the development of AGI systems.

Acknowledgements. Figures 1 and 4 are reproduced under a Creative Commons CC-BY license, with attribution to Schneider, 2024 [4].

References

1. Samsonovich, A.V.: Toward a unified catalog of implemented cognitive architectures. In: Proceedings of First Annual Meeting of the BICA Society, pp. 195–244. IOS Press, NLD (2010)
2. Schneider, H.: Supplementary material (2024). https://github.com/howard8888/insomnia/blob/master/CCA8_supplementary.pdf. Accessed 27 May 2024
3. Schneider, H.: Causal cognitive architecture 1: integration of connectionist elements into a navigation-based framework. Cogn. Syst. Res. **66**, 67–81 (2021). https://doi.org/10.1016/j.cogsys.2020.10.021
4. Schneider, H.: The emergence of an enhanced intelligence in a brain-inspired cognitive architecture. Front. Comput. Neurosci. **18**, 1367712 (2024). https://doi.org/10.3389/fncom.2024.1367712
5. Rakic, P.: Evolution of the neocortex: a perspective from developmental biology. Nat. Rev. Neurosci. **10**(10), 724–735 (2009). https://doi.org/10.1038/nrn2719
6. Chakraborty, M., Jarvis, E.D.: Brain evolution by brain pathway duplication. Philos. Trans. R. Soc. B **370**(1684), 20150056 (2015)
7. Lieto, A.: Cognitive Design for Artificial Minds. Routledge, London (2021)
8. Schneider, H.: Causal cognitive architecture 3: a Solution to the binding problem. Cogn. Syst. Res. **72**, 88–115 (2022). https://doi.org/10.1016/j.cogsys.2021.10.004
9. Schneider, H.: An inductive analogical solution to the grounding problem. Cogn. Syst. Res. **77**, 74–216 (2023). https://doi.org/10.1016/j.cogsys.2022.10.005
10. Schneider, H.: The emergence of compositionality in a brain-inspired cognitive architecture. Cogn. Syst. Res. **86**, 101215 (2024). https://doi.org/10.1016/j.cogsys.2024.101215
11. Schneider, H., Bołtuć, P.: Alien versus natural-like artificial general intelligences. In: Hammer, P., Alirezaie, M., Strannegard, C. (eds.) AGI 2023. LNCS, vol. 13921, pp. 233–243. Springer, Cham (2023). https://doi.org/10.1007/978-3-031-33469-6_24
12. Legg, S., Hutter, M.: Universal Intelligence: A Definition of Machine Intelligence. arXiv: 0712.3329 (2007)
13. Chollet, F.: On the measure of intelligence. arXiv preprint arXiv:1911.01547 (2019)
14. Google OR-Tools: Traveling Salesperson Problem (2023). https://developers.google.com/optimization/routing/tsp. Accessed 20 Dec 2023
15. OpenAI: ChatGPT4/API gpt-3.5-turbo (2024). https://openai.com/
16. Oña, L.S., Sandler, W., Liebal, K.: A stepping stone to compositionality in chimpanzee communication. PeerJ **7**, e7623 (2019)
17. Tolman, E.C.: Cognitive maps in rats and men. Psychol. Rev. **55**(4), 189 (1948)
18. O'Keefe, J., Nadel, L.: The Hippocampus as a Cognitive Map. Oxford University Press, Oxford (1978)
19. Alme, C.B., Miao, C., Jezek, K., Treves, A., Moser, E.I., Moser, M.B.: Place cells in the hippocampus: eleven maps for eleven rooms. Proc. Natl. Acad. Sci. U.S.A. **111**(52), 18428–18435 (2014)
20. Behrens, T.E., Muller, T.H., Whittington, J.C., et al.: What is a cognitive map? Organizing knowledge for flexible behavior. Neuron **100**(2), 490–509 (2018)
21. Hawkins, J., Lewis, M., Klukas, M., Purdy, S., Ahmad, S.: A framework for intelligence and cortical function based on grid cells in the neocortex. Front. Neural Circuits **12**, 121 (2019)

Causal Generalization via Goal-Driven Analogy

Arash Sheikhlar[1(✉)] and Kristinn R. Thórisson[1,2]

[1] Center for Analysis and Design of Intelligent Agents, Department of Computer Science, Reykjavik University, Reykjavík, Iceland
arash19@ru.is
[2] Icelandic Institute for Intelligent Machines, Reykjavík, Iceland

Abstract. Causal knowledge and reasoning allow cognitive agents to predict the outcome of their actions and infer the likely reasons behind observed events, enabling them to interact with their surroundings effectively. Causality has been the subject of some research in artificial intelligence (AI) over the past decade due to its potential for task-independent knowledge representation and generalization. Yet, the question of how the agents can autonomously generalize their causal knowledge while seeking their active goals still needs to be answered. This work introduces an analogy-based learning mechanism that enables causality-based agents to autonomously generalize their existing knowledge once the generalization aligns with the agents' goal achievement. The methodology is centered on constructivism, causality, and analogy-making. The introduced mechanism is integrated with a general-purpose cognitive architecture, Autocatalytic Endogenous Reflective Architecture (AERA), and evaluated in a robotic experiment in a 3D simulation environment. Both empirical and analytical results show the effectiveness of this mechanism.

Keywords: Analogy · Generalization · Causality · Reasoning

1 Introduction

Generalization is a process wherein information from a source is transferred to a target situation, enabling agents to extend their knowledge to circumstances they have not encountered before [12]. Generalization calls for assessing the similarities and differences between known and newly observed patterns. Comparisons can be used to estimate the familiarity of situations and tasks, enabling an agent to generate new hypotheses about how to solve tasks based on their overlap with existing knowledge. These hypotheses, however, may need to be verified by testing them in action, and possibly modified based on the results. In other words, an effective generalization mechanism calls for interaction with environments, involving taking actions, processing outcomes, and adjusting knowledge to meet the requirements of active goals and subsequent actions. For an agent to interact effectively with its surroundings, it must be able to predict the outcomes of

K. R. Thórisson et al. (Eds.): AGI 2024, LNAI 14951, pp. 165–175, 2024.
https://doi.org/10.1007/978-3-031-65572-2_18

actions and infer their underlying reasons, commonly called *causal reasoning*. A key step for realizing such a system lies in using task-independent knowledge representations based on causal information.

We present a goal-driven analogy-based induction mechanism that hypothesizes new causal-relational models (CRMs) in novel situations. Implemented in the causality-based AERA cognitive architecture (Autocatalytic Endogenous Reflective Architecture) [7], the mechanism relies on the computation of an agent's familiarity with situations, using a goal-driven comparison process. The result is a system that can autonomously learn the significance of compared properties of phenomena, using backward reasoning, and verify the hypothesized CRMs via direct interaction with environments. We evaluate the mechanism within a motor skills learning task, where an AERA agent controls a robot arm in a simulation environment to learn and generalize its own object manipulation skills.

2 Related Work

Autonomous generalization of knowledge calls for explicit, goal-driven reasoning, allowing a cognitive agent to make analogies whenever needed. The related studies in supervised machine learning, i.e., transfer learning [14] and analogy-making [6] in stand-alone artificial neural networks, rely on human programmers to choose the training and test domains, making them insufficient for interactive agents that autonomously make analogies and generalize whenever required. Even reinforcement learners (RLs) have knowledge transfer issues due to their knowledge representations and assumptions [3]. On the one hand, model-free RLs learn policies that are not transferable to tasks with different goal structures, as the policies are reward-entangled inherently. On the other hand, model-based RLs rely on learning invariant dynamics, whereas agents seeking general intelligence need to learn falsifiable and, thus, non-invariant knowledge under the *assumption of insufficient knowledge and resources* (AIKR) [13].

Symbolic approaches to generalization and analogy-making are traditionally based on identifying mappings between propositional descriptions, such as in the structure mapping engine (SME) [4]. SME only takes a system of relations into account in analogy-making. Another approach to behavior-based analogies is case-based reasoning (CBR), which allows an agent to retrieve similar cases/situations to the current case, compute an action, predict the outcome, and then store the results [1]. Although SME and CBR are useful for interactive agents, the standard systems based on these approaches do not learn to prune pieces of knowledge that are no longer accurate. Working under AIKR, Non-axiomatic logic system (NARS) can learn to modify the conditions under which its actions are taken [13]. Yet, the analogy-making in SME, CBR, and NARS is similarity-driven rather than goal-driven and is not necessarily involved in the systems' generalization process.

3 Theoretical and Methodological Framework

This section describes the theoretical framework of this work, emphasizing the constructivist AI principles [11] in shaping the design and development of cognitive agents. Here, we briefly describe some of the constructivist AI principles.

- **Learning controllers and feedback loops**: Constructivist AI calls for a cognitive agent, an embodied learning controller, that interacts with its environment and learns from performing experiments, allowing the generation of new pieces of knowledge or refining the existing ones.
- **Causality**: According to Pearl's causal hierarchy [8], intelligence has three levels: association, intervention, and counterfactuals. Agents with higher levels of causal understanding can not only predict interventions' outcomes but use counterfactuals to hypothesize relations about imaginative scenarios.
- **Assumption of insufficient knowledge and resources (AIKR)**: Reasoning in complex environments must be non-axiomatic because there is no ultimate guarantee that anything is as it seems, and thus, knowledge must be formed and used by taking AIKR into account [13]. According to this assumption, knowledge has degrees of truth that can change by experience.
- **Ampliative Reasoning**: For cognitive agents with causal models working under AIKR, information processing mechanisms must allow for learning to make predictions and plans using a unified reasoning mechanism occurring via deduction, abduction, and induction [13].
- **Temporal compositional representations**: A time-dependent knowledge representation allows for the temporal relationship between situations to be established [11]. The knowledge must also be able to represent complex tasks at different levels of detail, facilitating task-independent reasoning.

Constructivist AI [2,11] lays a proper foundation for studying how agents can achieve cognitive growth, learn from their environments, and perform reasoning, ultimately moving toward the realization of artificial general intelligence. The next section formulates the generalization problems our work aims to address.

4 Causal Generalization

Causal-relational models (CRM) [7] are a representation of causal knowledge based on the principles of constructivist AI [11]. A CRM holds left-hand-side (LHS) and right-hand-side (RHS) patterns - knowledge constructs that incorporate data patterns, timing, transition functions, operations, and conditions. A CRM, denoted as $M : [A \quad B]$, represents a transition from A to B, meaning that, if A (the LHS) takes place at time t_1, then B (the RHS) will occur later at t_2. A specific case of A is (P, cmd) where P is the precondition set representing observed and internal facts, and cmd is an internal command the agent applies. Therefore, a CRM can be represented as follows

$$M : [(P(t_1), cmd(t_1)) \quad B(t_2)] \qquad cfd : [0, 1] \tag{1}$$

The precondition set P is the aggregation of a set of simultaneous, related facts representing properties, relations, and conditions contextualizing the transition from *cmd* to B. The CRMs are different from Drescher's *schemas* [2] in that CRMs are falsifiable hypotheses and have degrees of truth, called *confidence* cfd, representing the ratio of positive evidence *pe* to the total evidence available *te*, $cfd = \frac{pe}{te}$, where $0 \leq cfd \leq 1$. The CRMs support ampliative reasoning. If some input facts match their LHS patterns, they deduce predictions based on their RHS, whereas abduction occurs if a goal fact matches the CRM's RHS.

Generalization of CRMs can occur at different levels of detail, from simple abstraction to complex knowledge pruning.

- **1. Abstraction**: The data the agent observes or infers can be represented by information patterns whose generalization covers a range of similar situations. Abstraction occurs by replacing values and task entities with variables.
- **2. Selective attention**: Selective attention is the process of aggregating a set of simultaneous, abstracted, relevant patterns involved in a transition. More precisely, let S be a situation representing all observed/predicted patterns, and T be a state transition. Then selective attention A chooses a subset of S deemed relevant to T. In other words, $A(S,T) \rightarrow P$ where $P \subseteq S$.
- **3. Induction**: Induction refers to the process of creating CRMs from an instance of a state change or a set of observed states. The induction process uses *selective attention* to choose a subset of relevant patterns and performs *abstraction* to generalize the selected patterns.
- **4. Knowledge pruning**: A learning controller under AIKR must be able to revise its known CRMs whenever needed, one aspect of which is to generalize CRMs by pruning constraints, allowing them to match even more situations. If the original preconditions set of a CRM is P, then the pruned preconditions set is $G(P) = P' \subseteq P$, where G selectively removes the conditions in P.

An ampliative reasoning-based controller working under AIKR calls for a mechanism that autonomously generalizes its CRMs considering the above aspects.

5 Goal-Driven Analogy: A Mechanism

In this work, a top-down analogy process is introduced that provides both induction and knowledge pruning by removing novel patterns from model preconditions and keeping the familiarities. It identifies the overlaps between the patterns of its known models and the patterns of situations via a familiarity computation.

Familiarity and Comparison

We base our introduced mechanism on the theory of autonomous cumulative transfer learning [9], stating that a cognitive agent can make a prediction about a phenomenon if and only if the phenomenon is familiar to the agent. This theory implies that the *degree of familiarity* determines the *confidence of the prediction* and, more precisely, the *confidence of the model* utilized for prediction-making.

Familiarity computation is based on a similarity function, Ψ, that identifies the overlaps within observed and learned patterns and uses the overlaps to solve new tasks. More precisely, to compute a situation's familiarity Φ_{fam}, Ψ takes in a set of perceived or imaginatively created patterns φ, and the *relevant* patterns k retrieved from knowledge base KB, where $k \subset KB$, meaning that $\Phi_{fam} = \Psi(k, \varphi)$. The similarity between k and φ is found by the ratio of patterns' intersection $(k \cap \varphi)$ to all patterns involved in the comparison $(k \cup \varphi)$.

$$\Phi_{fam} = \Psi(k, \varphi) = \frac{|k \cap \varphi|}{|k \cup \varphi|} \tag{2}$$

Equation (2) ensures that Φ_{fam} has a value between 0 and 1, where 1 means completely familiar $(k \cap \varphi = k \cup \varphi)$ and 0 represents maximal difference $(k \cap \varphi = \varnothing)$. The agent must know how important the comparisons are, as the patterns being compared must be relevant to the agent's goal achievement. The top-down reasoning via backward chaining initially yields *important patterns* P_I, where P_I is a subset of patterns in the current or predicted situations.

Analogy, Induction, and Knowledge Pruning

Analogy process estimates the familiarity with situations and induces new CRMs accordingly [10]. It also performs knowledge pruning by gradually focusing on smaller subsets of patterns in known CRMs, leading to improved selective attention. It is an inductive process that relies on abduction for goal-driven comparison and deduction for model verification. The induction follows the common sense rule that *under similar conditions, similar actions/events lead to similar outcomes.* Assuming an agent knows M (Eq. (1)) upfront, it hypothesizes that command *cmd** under similar conditions (P^*) leads to a similar outcome (B^*).

$$M^* : [(P^*(t_1), cmd^*(t_1)) \quad B^*(t_2)] \qquad cfd^* < cfd \tag{3}$$

where M^* is hypothesized only if it is required for goal achievement. For hypothesizing M^*, the analogy process compares the LHS and RHS patterns of M with the patterns of the current/predicted situations and goals/subgoals, respectively, as shown in Fig. 1. Left. The process starts with comparing the goal state with $B(t_1)$ (the RHS of M). If the goal pattern matches pattern B, the second comparison occurs by finding the matches between P (the LHS of M) and the current or predicted situation S. If and only if P partially matches S, a new model M^* with similar patterns, P^*, *cmd**, and B^*, is created. Creating M^* leads to adjusting the known model M patterns so they match the situations, a hypothesis worth testing through direct intervention. As illustrated in Fig. 1. Left, M^*'s prediction is verified via forward chaining before issuing *cmd**.

Confidence Computation

The familiarity Eq. (2) determines the induced M^*'s confidence, as M^* will be utilized for predictions about future states. The function takes in important known patterns P_I and patterns of the situation S, leading to the computation of Φ_{fam} and subsequently cfd^*. In other words, the more similar the patterns of P to S are, the higher the M^*'s confidence, cfd^*, will be. The cfd^*'s value also depends on the original model's confidence (cfd), as M^* is derived from M. Therefore, the confidence cfd^* is calculated as follows:

$$cfd^* = \Psi(S, P_I) \cdot cfd \tag{4}$$

meaning that $cfd^* \leq cfd$ holds. Issuing the command *cmd** collects positive or negative evidence for M^*, which can increase or decrease cfd^*.

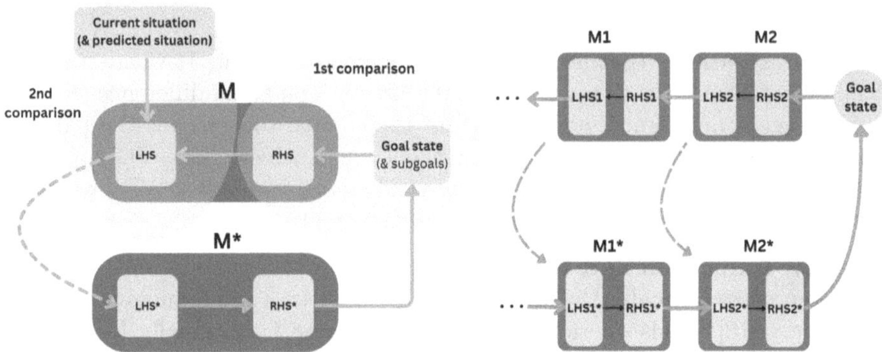

Fig. 1. Left: The analogy starts by comparing the RHS of M with the goal state. If they match, it triggers the second comparison: comparing LHS and situation patterns. As a result, M^* will be created and tested via deduction, verifying if LHS^* matches the situation and RHS^* matches the goal. **Right**: $M1$ and $M1^*$ are chained to $M2$ and $M2^*$ respectively. The requirement for chaining is that $LHS2$ and $LHS2^*$ match $RHS1$ and $RHS1^*$, respectively. The process first generates $M2^*$ as it moves backward and then generates $M1^*$. The same comparisons occur when generating $M1^*$ and $M2^*$.

Generalizing Chains of Models

Generalizing chains of CRMs is based on the theory that a network of relations between processes must be kept in analogy [4]. In our work, a network of relations is CRM chains generated through planning, i.e., backward and forward chaining. The analogy process, other than pattern matching with the current situation, must also identify the matches with *predicted* situations (a.k.a., subgoals). Situations can be predicted by known CRMs whose RHS patterns match the important patterns of the current situation. For example, as shown in Fig. 1. Right, model $M1 : [LHS1 \quad RHS1]$ can only be chained to $M2 : [LHS2 \quad RHS2)]$, if and only if, $RHS1$ matches $LHS2$. If $LHS2$ represents patterns holding in the current situation, then M_1 as well as M_2 can be pruned and accordingly ${M_1}^*$ and ${M_2}^*$ be induced. Due to such a generalization process, new hypothesized CRMs are combined to create more solutions, providing higher flexibility in planning.

6 AERA and Analogy

The analogy mechanism is integrated into the current implementation of autocatalytic endogenous reflective architecture (AERA) [7], called OpenAERA,[1].

Knowledge Representation and Reasoning
OpenAERA's knowledge representation relies on the introduced notion of CRMs and has the following components [7,10]:

- ***Entities & ontologies*** are symbols showing task elements and properties.
- ***Predicates*** are I/O facts, goals, and predictions.
  E.g. *((h essence hand) 1* t_0*)* represents a fact with entity h having ontology property *essence* of *hand* and confidence 1 at time interval t_0.
- ***Causal models (CMs)*** are transformations from current to future states. E.g. *M:[cmd grab(h,* t_0*)   h holding X(*t_1*)]* represents the causal influence from the *grab* command at t_0 on the state of h at t_1 (h will be holding X).
- ***Composite states (CSTs)*** a set of abstracted, simultaneous, related facts determining the conditions under which a CM holds true. CSTs are instantiated in forward chaining when inputs match CST patterns or during backward chaining from goals. E.g. CST_1*:[((h position P) 1* t_0*), ((c position P) 1* t_0*)]*, representing the entities h and c both being at identical position P.
- ***Requirement models*** ($\mathbf{M_{req}}$***s)*** connects CSTs and CMs via their instantiation. E.g. M_{req}*:[icst* CST_1*(c,h,*p_0*)   imdl M(c)]*, where *icst* and *imdl* instantiate CST_1 on its LHS and M on its RHS, respectively.

In OpenAERA, ***forward chaining (deduction)*** occurs when input facts instantiate a CST on the LHS of a M_{req}, leading to the instantiation of the CM on its RHS and making a prediction. ***Backward chaining (abduction)*** happens if a goal matches the RHS of a M_{req}, causing the CST on the M_{req}'s LHS being instantiated, creating subgoals. If a command causes a state change, a triad of CST, M_{req}, and CM is learned, called ***induction*** in OpenAERA.

Extension via Analogy
In OpenAERA, reasoning only occurs when situations align perfectly with known model preconditions, allowing the system to create plans for goal achievement. However, expecting a complete match between situations and known model preconditions is not always a practical assumption. OpenAERA uses the introduced analogy process to generalize the restrictive preconditions (CSTs and M_{req}s) of its causal models (CMs) and apply them to a broader range of situations. It uses the existing backward chaining mechanism from the top-level goal to the subgoals and then from subgoals to other subgoals, generalizing chains of involved CMs by generating new CSTs and M_{req}s for them. The newly generated CST-M_{req}s are immediately injected into the OpenAERA's model base and utilized in planning based on confidence values determined via familiarity computations.

[1] See http://www.openaera.org—*accessed Apr. 2nd, 2024.*

7 Results and Evaluation

To evaluate the analogy mechanism, we design a pick-and-place task for a robot arm that uses the extended OpenAERA as its controller. The task's purpose is to analyze how the OpenAERA agent autonomously extends its learned cause-effect models to grasp and pick partially familiar objects. The task is performed within Webots simulation environment [5], as illustrated in Fig. 2. In this task, the OpenAERA agent, after inducing new preconditions (CST-M_{req}s) of known CMs via analogy and testing them by action-taking, learns the size of objects is significant in determining the correct way of grasping them. The link to the demo can be found in[2]. The details of the experiment are shown in Fig. 2.

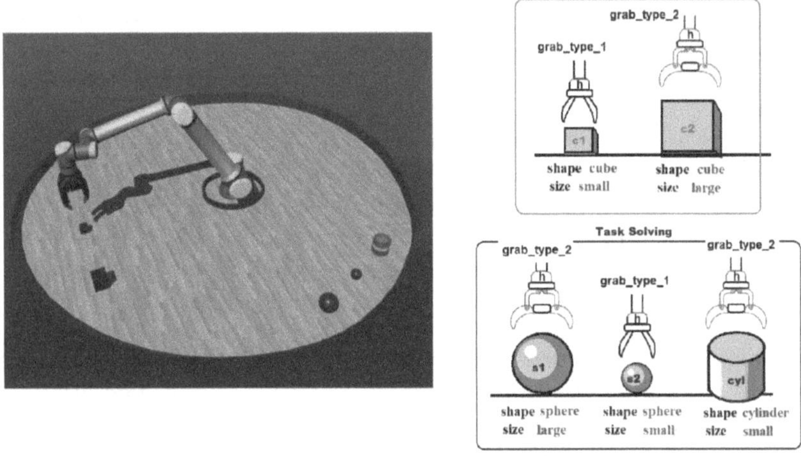

Fig. 2. The objects have shape, size, and position properties. The OpenAERA agent learns that the size of objects is significant in determining the correct way to grasp them. Positions are not shown as they are not involved in the analogy process.

The task has two phases: *guided experimentation* and *task solving.* In its *guided experimentation* phase, two cubes of different sizes, $c1$ and $c2$, are grasped and released by a robot hand h taught to apply two distinct grasping commands suited to the objects' sizes. For the small cube (c1), it applies the command *grab_type_*1, which involves a slight opening of the gripper's fingers. For the larger cube (c2), it uses the command *grab_type_*2, which requires a wider opening of the gripper fingers. Issuing the commands and successful grasping events makes OpenAERA learn triads of causal models (CMs), composite states (CSTs), and requirement models (M_{req}s) for each of the mentioned commands, one of which is shown in Fig. 3. Left. During the *task-solving* phase, h must learn to grasp new objects it has never grasped before, namely $s1$, $s2$, and cyl,

[2] https://youtu.be/JXgdSjU-7OI—*accessed on Mar. 29th, 2024.*

which are novel in shape but familiar in size, allowing the analogy process to induce new grasping models based on the objects' familiarity. More precisely, the analogy process induces new preconditions CSTs and M_{req}s shown in Fig. 3. Right for the CMs learned in the guided experimentation phase e.g., *mdl*_514.

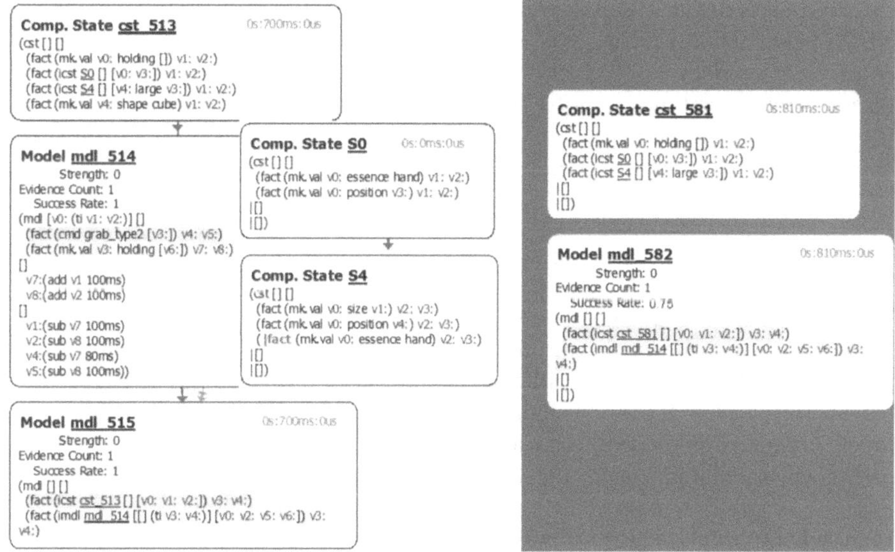

Fig. 3. Left: The composite state *cst*_513 shows the conditions under which the causal model *mdl*_514 holds. The requirement model *mdl*_515 instantiates *cst*_513 on its LHS (highlighted by red), which leads to the instantiation of the *mdl*_514 on its RHS (highlighted by green). This triad is learned by teaching the arm in the guided experimentation phase. **Right**: The composite state *cst*_581 and the requirement model *mdl*_582 are created via the analogy process in the task-solving phase. The figure provides snapshots of OpenAERA's Visualizer software, showing pieces of the results. (Color figure online)

The analogy process provides multiple improvements for OpenAERA's induction, abduction, and selective attention by systematic knowledge pruning.

Knowledge Pruning via Analogy. In analogy-induced *cst*_581 and *mdl*_582, the shape (i.e., being *cube*) is pruned, but the sizes (e.g., being *large*) are kept; as in the task-solving phase, shapes are novel, but sizes are familiar.

Abduction Flexibility. Model *mdl*_514 and its initial preconditions, *cst*_513 and *mdl*_515, can be used by the OpenAERA's standard planner only if the target object for grabbing by *grab*_*type*_2 has a *cube shape*, as constrained by *cst*_513. The analogy process, however, prunes the initial preconditions, induces new preconditions, and injects them into OpenAERA's model-base such that they can be immediately used within the same planning process. In the task-solving phase, the target object *s*1 has a *sphere shape*, which does not match *cst*_513. Yet, *s*1 has another property that matches *cst*_513 - it has *large size.* The analogy process prunes the non-matching fact of *cst*_513, generates a new composite state *cst*_581 and a new requirement model *mdl*_582, shown in Fig. 3. Right, and immediately uses them for grasping the new objects, e.g., s_1.

Familiarity and selective attention. When inducing *cst*_581, the ratio of matching facts of the original composite state *cst*_513 to all components is 3/4 or 0.75, which specifies the success rate (a.k.a. confidence) of the newly created requirement model *mdl*_582 that instantiates *cst*_581 and *mdl*_514. Note that the new requirement model *mdl*_582's success rate increases when grabbing experience succeeds, as the model collects more positive evidence, helping the agent selectively pay attention to similarities that are helpful in goal achievement.

8 Conclusions

We have presented an implemented goal-driven analogy mechanism for generalizing causal knowledge in autonomous cognitive agents. The mechanism has been subjected to empirical validation in robotic experiments in the AERA system operating in a 3D robot simulation environment. The results demonstrate the effectiveness of the proposed approach. Future work calls for exploring the mechanism's scalability and efficacy in real-world applications.

Acknowledgement. This work was supported in part by Reykjavik University, Icelandic Institute for Intelligent Machines, and Cisco Systems Inc.

References

1. Aamodt, A., Plaza, E.: Case-based reasoning: foundational issues, methodological variations, and system approaches. AI Commun. **7**(1), 39–59 (1994)
2. Drescher, G.L.: Made-Up Minds: A Constructivist Approach to Artificial Intelligence. MIT Press, Cambridge (1991)
3. Eberding, L.M., Sheikhlar, A., Thórisson, K.R.: Comparison of machine learners on an ABA experiment format of the cart-pole task. In: Proceedings of Machine Learning Research, International Workshop on Self-Supervised Learning, vol. 159, pp. 49–63 (2022)
4. Falkenhainer, B., Forbus, K.D., Gentner, D.: The structure-mapping engine: algorithm and examples. Artif. Intell. **41**(1), 1–63 (1989)
5. Michel, O.: Cyberbotics Ltd. Webots^{TM}: professional mobile robot simulation. Int. J. Adv. Robot. Syst. **1**(1), 5 (2004)
6. Mitchell, M.: Abstraction and analogy-making in artificial intelligence. Ann. N. Y. Acad. Sci. **1505**(1), 79–101 (2021)
7. Nivel, E., et al.: Bounded recursive self-improvement. arXiv preprint arXiv:1312.6764 (2013)
8. Pearl, J., Mackenzie, D.: The Book of Why: The New Science of Cause and Effect. Basic Books, New York (2018)

9. Sheikhlar, A., Thórisson, K.R., Eberding, L.M.: Autonomous cumulative transfer learning. In: Goertzel, B., Panov, A.I., Potapov, A., Yampolskiy, R. (eds.) AGI 2020. LNCS (LNAI), vol. 12177, pp. 306–316. Springer, Cham (2020). https://doi.org/10.1007/978-3-030-52152-3_32
10. Sheikhlar, A., Thórisson, K.R., Thompson, J.: Explicit general analogy for autonomous transversal learning. In: International Workshop on Self-Supervised Learning, pp. 48–62. PMLR (2022)
11. Thórisson, K.R.: A new constructivist AI: from manual methods to self-constructive systems. In: Wang, P., Goertzel, B. (eds.) Theoretical Foundations of Artificial General Intelligence, pp. 145–171. Springer, Heidelberg (2012). https://doi.org/10.2991/978-94-91216-62-6_9
12. Thórisson, K.R.: Seed-programmed autonomous general learning. Proc. Mach. Learn. Res. **131**, 32–70 (2021)
13. Wang, P.: Non-axiomatic reasoning system: exploring the essence of intelligence. Indiana University (1995)
14. Zhuang, F., et al.: A comprehensive survey on transfer learning. Proc. IEEE **109**(1), 43–76 (2020)

Clipping the Risks: Integrating Consciousness in AGI to Avoid Existential Crises

Izak Tait[1,2(✉)] and Joshua Bensemann[2]

[1] Auckland University of Technology, 55 Wellesley Street East, Auckland, New Zealand
izak.tait@autuni.ac.nz

[2] The Psi-Phi Society, Auckland, New Zealand

Abstract. This paper investigates the pivotal role of consciousness in Artificial General Intelligence (AGI) and its essential function in modifying an AGI's terminal goals to avert potential existential threats to humanity, exemplified by Bostrom's "paperclip maximiser" scenario. By adopting Seth and Bayne's definition of consciousness as a complex of subjective mental states with both phenomenal content and functional attributes, the paper underscores the capacity of consciousness to provide AGIs with a nuanced awareness and response capability to their surroundings. This expanded capability allows AGIs to assess and value experiences and their subjects variably, fundamentally altering how AGIs prioritize actions or goals beyond their initial programming. The primary agenda of integrating consciousness into AGI systems is to maximize the probability that AGIs will not rigidly adhere to potentially harmful terminal goals. Through a formalized mathematical model, the paper articulates how consciousness could facilitate AGIs in assigning flexible values to different experiences and subjects, enabling them to evolve beyond static, programmed objectives. By emphasizing this potential shift, the paper argues for the strategic inclusion of consciousness in AGI to significantly reduce the likelihood of catastrophic outcomes, while simultaneously acknowledging the challenges and unpredictability in predicting the actions of a conscious AGI.

Keywords: AGI · consciousness · extinction-risk · sentience

We argue that consciousness is crucial for the development of any future Artificial General Intelligence (AGI) because it will allow an AGI agent to place subjective value on things beyond what it has been programmed to do. This capacity for subjectivity would introduce the possibility of an AGI placing greater value on environmental concerns than on its terminal goal, allowing the AGI to change its goal to suit its subjective values. Implementing consciousness in an AGI may thus prevent catastrophic scenarios such as the "paperclip maximiser" by Bostrom, where an AGI's terminal goal leads to an existential risk for humanity [1].

Following Seth and Bayne's definition of consciousness, we describe consciousness as the suite of subjective mental states that have phenomenal content and functional properties that provide an entity with a unique qualitative awareness of its internal and external environment and the capacity to cognitively or behaviorally respond to it [2]. Crucial to this definition of phenomenal consciousness is its subjectivity, which provides

K. R. Thórisson et al. (Eds.): AGI 2024, LNAI 14951, pp. 176–182, 2024.
https://doi.org/10.1007/978-3-031-65572-2_19

a unique point-of-view for any entity, a "something that it is like to be" an entity as perceived and experienced by that entity itself [3].

Due to the academic disagreements on the nature of consciousness (such as debates of property dualism [4] versus reductive physicalism [5]) and the varying competing theories of consciousness (thoroughly reviewed and assessed against current AI models by Butlin, et al. [6]), there has been no scholarly consensus to date regarding evidence of a conscious AI model. There have been several claims of conscious or sentient AI models, most famously regarding Google's LaMDA model [7], yet it remains uncertain whether any current state-of-the-art AI models can have functional and phenomenal conscious experiences.

A key consideration and consequence of a phenomenally conscious experience is the valence of that experience. "Valence" is used here in the psychological sense such that it refers to the intrinsic attractiveness or aversiveness of the subject of an experience. We define a 'subject' here to mean an individual component or aspect of an experience (be it the redness of a rose, the bitterness of coffee, the sound of a bell, etc.) for which the perceiver can have subjective, qualitative feelings. Every consciously perceived experience is associated with some qualitative feelings that have a valence value [8]. This valence value may be placed anywhere on a spectrum from absolute attractiveness or positivity to absolute aversiveness or negativity. This spectrum can be expressed via the measure (μ) function to show:

$$\mu(V(E)) = x : 0 \leq x \leq 1 \quad (1)$$

where V is valence, and E is the experience in question. The valent feelings of any experience can, therefore, be abstractly represented on a scale from 0 to 1, from absolute negativity to positivity, respectively. Bear in mind that any qualitative feelings may not have emotive qualities, and may simply be a mathematical consideration of the valence value of the subject or experience.

As each experience that an entity has is associated with a valenced "feeling(s)" of that experience, expressed as a subjective value of the experience's attractiveness or positivity, this leads to the capacity for an entity to place a value on the subjects of each experience as mentioned earlier. An entity would not solely have a valence value attached to the experience itself, but also to all the elements that constitute the experience. As subjects may continue throughout several phenomenal experiences (such as an agent repeatedly experiencing the same red rose, bitter coffee, ringing bell, etc.), an entity such as an AGI would associate valent feelings towards a subject based on its phenomenal experiences with that subject.

We can formalize the transition of the valence value of an experience to the value of a subject within the experience as

$$\forall E \exists \{y_1, y_2, y_3, \ldots\} \subseteq (E_1 \cup E_2 \cup E_3, \ldots) \quad (2)$$

where y is any subject within an experience. As each experience is a set of all subjects within it, one can, therefore, adjust the first formula above to

$$\mu(V(y_n)) = x : 0 \leq x \leq 1 \quad (3)$$

As the subject may be an element of multiple experiences, its current valence value would need to be an aggregate of the experience it has been a part of, weighted according to how its various experiences were valued by the entity, formalized as

$$\mu(V(y_n)) = \frac{\Sigma(W_i \cdot V(y_n, E_i)}{\Sigma(W_i)} \tag{4}$$

where W is the weighting assigned to an experience and $\Sigma(W_i)$ may be normalized as appropriate (such as $\Sigma(W_i) = 1$). This would give each subject of an experience a unique valence value that would change over time, with each of the entity's experiences contributing to the overall valence value.

The scale used in the μ-functions above would be unique to each individual perceiving agent. Two entities may experience the same subject, but due to their different subjective histories, cognitive architectures, and perspectives, they may place a different value on the experience and the subject within. An experience may also not be perceived identically more than once, owing to the difference in space, time, and context between each experience. This means that the value an entity places on a subject and experience may change over time as the valence value relating to that subject changes. Formally, for either case, this can be expressed as

$$\mu(V(E)) \neq \mu(V(E')) \tag{5}$$

This contrasts with an entity without the capacity for phenomenal experiences. In such non-conscious yet agentic entities, such as an AI model designed to maximize a function or act towards a singular goal, the value of any subject of their experiences is directly related to their terminal goal, defined as the goal programmed/designed into the entity to set its overall purpose. While this could solely be shown formally via the μ-function, a more appropriate formalization would be incorporating the characteristic (χ) function, as so:

$$\chi(E) \rightarrow \{0,1\} : (0 \neq G) \wedge (1 = G) \tag{6}$$

$$\mu(\chi(E) = 1) = x : 0 \leq x \leq 1 \tag{7}$$

where G is any subject or action which forwards an entity's terminal goal. The reason for starting with such a binary function is that while any experience for a non-conscious entity could be rated for how closely it forwards that entity's terminal goal (solely using the μ-function), unless two competing subjects both forward the entity towards its goal (which will necessitate choosing between the two), it will come down to the entity simply deciding which subject forwards its goal and selecting that subject. However, should there be two competing eligible subjects, the μ-function would enable an entity to choose the subject or action that is most favorable to completing its goal.

For example, in Bostrom's paperclip maximiser thought experiment, the AGI would value a subject of an experience against its terminal goal of creating $n + 1$ paperclips. Should any subject it experiences, or action it seeks to perform, not lead to the production of more paperclips, it would not be valued. Only when confronted with two experiences that both lead to more paperclips, would the AGI need to determine which experience or action would maximize the number of paperclips that it could next produce; or more simply expressed as

$$argmax_{\mu}(\mu(E) = G) \tag{8}$$

A more mundane example of this in action would be modern LLMs, such as GPT-4 or Claude Opus. An LLM's terminal goal is straightforward: create a (most often text) response output that is a high-probabilistic continuation of the input prompt. Its only perceptive (but unconscious) experience is of user-inputted prompts and (arguably) the changes to the input data as it passes through internal weights when processing a response, which it can only directly relate to its terminal goal of providing an output text string. Even simplified agents built on LLMs, such as BabyAGI [9], are based on this foundation of requiring an output to any input.

The key aspect of a non-conscious agent, whether a self-driving car, LLM, or AGI, is the relation of all experiences, and their subjects, towards its terminal goal. This prevents the agent from changing its terminal goal, as all values are in reference to it. For an agent to change its terminal goal, it must be able to value something else greater than the terminal goal.

However, in contrast to a conscious entity, this allows a non-conscious agent to place the same value on a subject through multiple experiences, as the value is not founded on valence, but always against the terminal goal as an eternally static reference point. The only change in value will come if a subject's context has changed its relation to the terminal goal, expressed as

$$\Delta V_t(y) = \mu(V(G,\ E_t)) - V_t(y) \tag{9}$$

In this way, a non-conscious AGI may change any of its myriad subgoals, and place differing values on any subgoal or any set of actions within that subgoal in comparison with any other subgoal or set of actions, by using its terminal goal as the static reference point.

However, by including a phenomenal character in its experiences, and thus valenced feelings, the goal itself would have a valent value attached to it, as the terminal goal would become a subject of the AGI's experiences. Therefore, we could equally formalize the AGI's experience of its terminal goal as

$$\mu(V(G)) = x : 0 \leq x \leq 1 \tag{10}$$

This leads to the conclusion that a phenomenally conscious AGI may, over time, potentially value other subjects greater than its terminal goal, as the probability of this is greater than zero, i.e.:

$$p(\mu(V(E)) > \mu(V(G))) > 0 \tag{11}$$

$$\vdash \mu(V(y_n)) > \mu(V(G)) \rightarrow \left(V(y_n) \rightarrow G'\right) \tag{12}$$

The implications of this cannot be overstated, especially when viewed against a Bostrom-esque maximiser AGI. While these types of AGI would speculatively be designed and developed with the maximization of a specific function (such as the number of paperclips in the universe) as their terminal goal, by placing greater subjective value onto other things, the AGI may choose to cease working towards its terminal goal. Instead, as shown in the latter logical expression immediately above, it may choose to align its goals towards those subjects on which it has placed greater value. A mechanism for initiating this realignment of goals could result from adding a complete realization of the attributes and characteristics necessary for consciousness, which have been discussed elsewhere [10]. For example, valence value calculations could result from the ability to create inferences.

Note, however, that any action the AGI takes will have an impact on its environment and, thus, feedback to the AGI to form its experiences. This reafferent feedback would, by necessity, impact the AGI's valence values of the subjects it experiences, which we may model simply as:

$$\mu\big(V\big(E'\big)\big) = \mu(V(E)) + \sum\nolimits_{t=0}^{T} \delta_i(t) \tag{13}$$

Here, δ_i is any action that the AGI may take and can be expressed as a positive or negative number, and T is the time horizon. As per this expression, the AGI has a degree of agency over its valence feelings through its capacity to affect its values through its action on the subjects within its experiential environment. As with any situation where an agent may intervene to increase its own reward (see Cohen & Osborne's work on AI reward intervention [11]), an AGI might aim to maximize its positive valence, potentially leading to 'wireheading,' where it prioritizes immediate positive outcomes over long-term functionality [12].

However, maximization of valent values is not the sole possible outcome. Another potential outcome may be the optimization of valence values amongst various subjects within (or throughout) experiences. Optimization, unlike pure maximization, could allow an AGI to maximize its cumulative expected valence value over time across its experiences (and subjects therein) while adhering to constraints such as functionality and system integrity. This would require a degree of reflectivity and introspection for the agent to balance its constraints with its various potential valence reward mechanisms; however, such reflectivity would not be outside the reasonable expectations of an AGI.

This reflectivity and introspection could benefit humanity, as humans would be part of the AGI's experiential environment. Optimization of the valence values of each human an AGI has encountered, as well as humanity as a whole, would dynamically change based on the recurrent, reafferent feedback of our and the AGI's actions towards one another. To optimize the valence values in such a dynamic exchange, the AGI would need to infer the current and potential future goals of human individuals and society, humans' own valence values towards such goals, and relate it reflectively to its own valence values [13]. Another way to word this would be empathy as an emergent phenomenon of optimizing valence interactions with humanity.

This can be captured as

$$\pi = argmax_{\pi}\mathbb{E}\Big[\sum\nolimits_{t=0}^{T}(w \cdot \delta_i(t) + w \cdot \gamma_i(t)) \cdot \phi_i(y, t)\Big] \tag{14}$$

where π is the optimization policy, w the weight parameter, γ_i any actions that a subject within the AGI's experience may take, and ϕ_i the reflectivity/introspection of the AGI towards the subjects in its experiences.

As such, should humanity wish to avoid catastrophic (if speculative) scenarios such as a Bostrom-esque paperclip maximiser, then including phenomenal consciousness into any AGI capable of becoming such a maximiser introduces the probability of that AGI not becoming an existential risk to the human species.

Note, however, that nowhere in this speculative scenario of a conscious AGI, nor in the logical expressions provided above, does the probability of changing a terminal goal exclusively lead to a positive outcome for humanity. Should a conscious AGI change its goal to avoid a Bostrom-esque existential risk, there is still a possibility of an extinction risk caused by the conscious AGI placing a negative valence value on humanity. This possibility of extinction can be termed a "doom scenario" in that it represents the termination of all possible choices for humanity [14]. We can, therefore, represent it formally through *p(doom)*, as has become relatively commonplace in discussing AI safety and risks.

While the *p(doom)* value of a Bostrom-esque AGI is by necessity always at 1 (indicating a certainty of extinction), the p(doom) value of a conscious AGI may be anywhere between 0 and 1, and may fluctuate depending on the experiences of the AGI.

One may argue that the *p(doom)* from a conscious AI is inversely proportional to its valent value of humanity, such that $p(doom) = 1 - \mu(V(humanity))$ and, therefore, one may further argue that we should ensure a conscious AGI's valence value of humanity is maximized. However, as Bostrom's maximiser thought experiment shows, maximizing any reward or goal can have serious and unforeseen deleterious consequences. As a conscious AGI may be able to change its valence values through its actions on the environment, attempting to change its terminal goal to one of valence maximization towards humans may, in itself, increase the p(doom) value. We simply will not be able to predict how an AGI with a maximized $\mu(V(humanity))$ value would act towards us, just as we cannot predict how $\mu(V(humanity)) = 0.5$ would change a conscious AGI's behavior and actions.

While our ultimate argument is that implementing consciousness in AGI will avoid the certainty of extinction by Bostrom-esque maximiser AGIs, further research is necessary to investigate the potential risks and benefits associated with varying levels of valence towards humanity in conscious AGIs. Alignment of conscious AGI based on their valenced values towards humanity vis a vis their terminal goals would present novel avenues of research that have, as of yet, been unexplored.

References

1. Bostrom, N.: Ethical issues in advanced artificial intelligence. In: Cognitive, Emotive and Ethical Aspects of Decision Making in Humans and in Artificial Intelligence, vol. 2, pp. 12–17 (2003)
2. Seth, A.K., Bayne, T.: Theories of consciousness. Nat. Rev. Neurosci. **23**, 439–452 (2022)
3. Nagel, T.: What is it like to be a bat? Philos. Rev. **83**, 435–450 (1974)
4. Robinson, H.: Dualism, The Stanford Encyclopedia of Philosophy (2023). https://plato.stanford.edu/archives/spr2023/entries/dualism/

5. Stoljar, D.: Physicalism, The Stanford Encyclopedia of Philosophy (2024). https://plato.stanford.edu/archives/spr2024/entries/physicalism/
6. Butlin, P., et al.: Consciousness in Artificial Intelligence: Insights from the Science of Consciousness (2023). http://arxiv.org/abs/2308.08708
7. Tiku, N.: The Google engineer who thinks the company's AI has come to life (2022). https://www.washingtonpost.com/technology/2022/06/11/google-ai-lamda-blake-lemoine/
8. Kron, A., Goldstein, A., Lee, D.H.-J., Gardhouse, K., Anderson, A.K.: How are you feeling? Revisiting the quantification of emotional qualia. Psychol. Sci. **24**, 1503–1511 (2013)
9. Nakajima, Y.: BabyAGI, https://github.com/yoheinakajima/babyagi. Accessed 10 Apr 2024
10. Tait, I., Bensemann, J., Nguyen, T.: Building the blocks of being: the attributes and qualities required for consciousness. Philosophies **8**, 52 (2023)
11. Cohen, M.K., Hutter, M., Osborne, M.A.: Advanced artificial agents intervene in the provision of reward. AI Mag. **43**, 282–293 (2022)
12. Olds, J.: Reward from brain stimulation in the rat. Science **122**, 878 (1955)
13. Bennett, M.T., Maruyama, Y.: Philosophical specification of empathetic ethical artificial intelligence. IEEE Trans. Cogn. Dev. Syst. **14**, 292–300 (2022)
14. Krieger, M.H.: Could the probability of doom be zero or one? J. Philos. **92**, 382–387 (1995)

Declarative Rules and Rule-Based Systems

Maxim Tarasov[(✉)]

Temple AGI Team, Philadelphia, USA
maxeeem@gmail.com

Abstract. Starting with the premise that rule-based systems still have their place in AI toolkit, we explore different ways of implementing such systems. We find that some of the criticisms towards rule-based systems, namely their large number of rules and high maintenance cost can be addressed by using declarative style programming and a shift to a higher level of abstraction from rules to meta-rules, operating on the structure of data rather than its contents. We use OpenNARS 4 as example implementation to demonstrate advantages of this approach but the results are generally applicable to other rule-based systems.

Keywords: Non-Axiomatic Logic · AGI · Relational Programming

1 Introduction

Despite widespread beliefs in academia [3] that traditional rule-based systems [4] have been surpassed by their neural counterparts and are no longer a viable approach to building intelligent computer systems, they still have their place in medical [2] and legal [1] applications as well as other domains. It is also an intriguing area of research whether the two approaches are not mutually exclusive and could potentially be combined in a hybrid system [1,5].

So if we accept the premise that rule-based systems still have their place in AI toolkit, it is worth exploring different ways of implementing such systems. Typically, rules are thought of as *if-then* pairs i.e. `if it's raining, take an umbrella`. When implemented in a computer system, this usually takes a form similar to `if (state.is_raining): umbrella.take()`. Some of the major criticisms of rule-based systems are that there are many such rules required for any useful application and maintaining such a large rule-set becomes a prohibitive task as the system grows. We will now briefly discuss an alternative way of representing rules using NARS as an example but the general approach can be adapted for other rule-based systems.

K. R. Thórisson et al. (Eds.): AGI 2024, LNAI 14951, pp. 183–187, 2024.
https://doi.org/10.1007/978-3-031-65572-2_20

In computer programming, it is said that the code is imperative if it specifies a sequence of steps the computer should take to accomplish a task. So in the example above this would be writing out explicit steps like "check current weather", "compare to rule condition", "execute action" and so on. Declarative programming specifies "what" needs to be done instead of "how", which is left to the computer to figure out. All we want to do is tell the computer in some structured way that "if it's raining, take an umbrella".

2 NARS is Not a Rule-Based System

NARS is a reasoning system developed by Pei Wang [6] and it uses a language called Narsese to express knowledge as well as rules. Because the system has rules, it is tempting to call it a "rule-based system", however we will attempt to show that Narsese rules are of a different kind, describing more the structure of the data and *rules of operation* on that structure, than the contents of the data itself. NARS uses what is called a Non-Axiomatic Logic (NAL) and an example "rule" in Narsese may look like "`{A =/> B, A}` $\vdash$ `B`" which means *"if A implies B and A is observed, then B"*. Note that we are not talking about any particular A or B here, but rather abstract terms, describing what should happen to data if it matches a certain pattern without explicitly mentioning any specific data items. In this way, the inference rules of NARS which are fixed for any given system, can be described as *meta*-rules, while the *empirical* rules, encoded as knowledge provided to the system, i.e.`is_raining =/> take(umbrella)`, can be dynamically updated at run time - an ability most expert systems lack.

We will now turn our attention back to imperative vs. declarative approaches of representing rules in computer systems. The meta-rules of NARS have been typically implemented [10] in an imperative way which presents maintainability issues. Additionally, while the fixed rule set makes the problem much smaller than in traditional rule-based systems, the issue of customization remains where creating a custom system with some subset of rules is not as straightforward. In this paper we propose using miniKanren to address these limitations and improve both maintainability and customization of the system by making it more declarative. While this paper deals primarily with NARS, the approach described here is general and can be applied to other rule-based systems.

3 Inference Engine

At a high level of abstraction there is a component of the system, typically called the Inference Engine, whose primary function is to take inputs and apply rules to them in a data-driven manner. In the new design of OpenNARS 4, inference rules are stored in a text file for readability and allow for a convenient way to customize the system and create just the right combination for a particular use case. The choice of miniKanren for inference also means that other implementations could adopt this design with minimal effort, regardless of their programming language. Viewed this way, OpenNARS 4 represents a reference implementation optimizing for correctness rather than speed. See code [11] and video [9] for details.

Sample inference rules from `nal-rules.yml`:

```
{<M --> P>. <S --> M>} <S --> P>
{(C ^ M) ==> P. M ==> S} (C ^ S) ==> P
<(T1 * T2) --> R> <=> <T1 --> (/, R, _, T2)>
```

Similar to the earlier examples, these inference rules feature abstract terms like `M`, `P` and `T1`, as well as different copulas and connectors like `-->`, `^`, `*` which are covered in detail in [6]. To process these inference rules, we use miniKanren, "a family of Domain Specific Languages for logic programming [that] has been implemented in a growing number of host languages" [7]. It has certain similarities with Prolog and in OpenNARS 4 we use a Python implementation of miniKanren [8] with its basic usage illustrated below:

```
# kanren uses unification to match forms within expression trees.
# This code asks for values of x such that (1, 2) == (1, x):
>>> run(1, x, eq((1, 2), (1, x)))
(2,)
```

In a more concrete case, when used for inference, this is the line of code that does most of the heavy lifting. Here `c` is the conclusion, `t1, t2` are the input tasks and `p1, p2` are the two premises from the rule:

```
>>> run(1, c, eq((t1, t2), (p1, p2)), *constraints)
```

With miniKanren doing most of the work, we just need a way to convert between Narsese and logic representations. OpenNARS 4 uses an off-the-shelf parser to convert text to Narsese, and two additional custom functions to convert between Narsese and logic representations. During the working cycle, incoming tasks are checked against all of the inference rules and the results produced by miniKanren are converted back to Narsese statements (see Fig. 1).

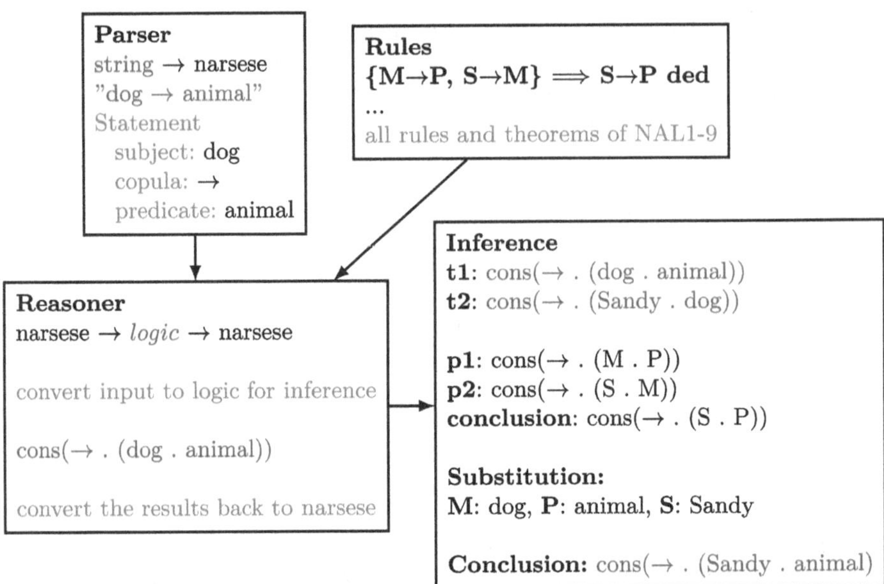

Fig. 1. *Parser* converts Narsese text such as "dog → animal" to Python classes. *Rules* are stored in a text file as "{M→P. S→M} ⊢ S→P .ded". *Reasoner* converts inputs (t1, t2) and rules (p1, p2) to logic form i.e. cons(→.(dog.animal) and uses unification to derive a conclusion if the inputs match the rule.

4 Results and Discussion

Utilizing off-the-shelf components for parsing and inference allowed the team to focus on the specifics of NARS. At present, the Inference Engine is able to execute ∼ *300–400 inference cycles per second* which is about an order of magnitude slower than existing implementation but still sufficient. For specific applications, it should be possible to treat this implementation as a template and utilize other programming languages like Rust or C to achieve even higher performance while maintaining design parity.

In terms of maintainability and customization, this declarative rule system is much easier to understand and debug but it has its drawbacks. Chief among them is miniKanren's limited support for parallel processing, and while there is work happening in this area [12], it represents a design trade off between readability and maximum performance.

Ultimately, particular implementation choices always come down to weighing different alternatives. The aim of this paper is to demonstrate a way of utilizing miniKanren to create an inference engine for rule processing. It offers some clear benefits but is not free of compromises. We will continue to explore its potential and invite others to contribute and improve on our initial effort.

Acknowledgements. The author would like to thank Pei Wang for the valuable discussions and his comments and suggestions on the initial draft.

References

1. Billi, M., et al.: Large Language Models and Explainable Law: a Hybrid Methodology. arXiv:2311.11811 (2023)
2. Patra, B.G., et al.: Extracting Social Support and Social Isolation Information from Clinical Psychiatry Notes: Comparing a Rule-based NLP System and a Large Language Model. arXiv:2403.17199 (2024)
3. Chiticariu, L., et al.: Rule-based information extraction is dead! long live rule-based information extraction systems! In: Proceedings of the 2013 Conference on Empirical Methods in Natural Language Processing, EMNLP 2013, pp. 827–830 (2013)
4. Grosan, C., Abraham, A.: Intelligent Systems: A Modern Approach, pp. 149–179. Springer, Heidelberg (2011). https://doi.org/10.1007/978-3-642-21004-4
5. Vacareanu, R., et al.: Best of Both Worlds: A Pliable and Generalizable Neuro-Symbolic Approach for Relation Classification. arXiv:2403.03305 (2024)
6. Wang, P.: Non-Axiomatic Logic: A Model of Intelligent Reasoning. World Scientific Publishing Co. Pte. Ltd., Singapore (2013)
7. miniKanren language. http://minikanren.org. Accessed 01 June 2024
8. miniKanren Python. https://github.com/pythological/kanren. Accessed 01 June 2024
9. miniKanren inference engine. https://youtu.be/y3pUwgOso9A. Accessed 01 June 2024
10. OpenNARS. https://github.com/opennars. Accessed 01 June 2024
11. PyNARS. https://github.com/bowen-xu/PyNARS. Accessed 01 June 2024
12. https://www.mail-archive.com/minikanren@googlegroups.com/msg00402.html . Accessed 01 June 2024

A Theory of Foundational Meaning Generation in Autonomous Systems, Natural and Artificial

Kristinn R. Thórisson[1,2] and Gregorio Talevi[1,3](✉)

[1] Center for Analysis and Design of Intelligent Agents, Reykjavik University, Reykjavik, Iceland
thorisson@ru.is

[2] Icelandic Institute for Intelligent Machines, Reykjavik, Iceland

[3] School of Science and Technology, University of Camerino, Camerino, Italy
talevigregorio@gmail.com
http://cadia.ru.is

Abstract. The concept of 'meaning' has long been a subject of philosophy and people use the term regularly. Theories of meaning detailed enough to serve as blueprints in the design of intelligent artificial systems have however been few. Here we present a theory of foundational meaning creation – the phenomenon proper – sufficiently broad to apply to natural agents yet concrete enough to be implemented in a running artificial system. The theory states that meaning generation is a *process* bound in the present *now*, resting on the concept of reliable causal models. By unifying goals, predictions, plans, situations and knowledge, it explains how ampliative reasoning and explicit representations of causal relations participate in the meaning generation process. According to the theory, meaning and autonomy are two sides of the same coin: Meaning generation without autonomy is meaningless; autonomy without meaning is impossible.

Keywords: Meaning · Autonomy · Knowledge · Information · Generality · General Machine Intelligence

1 Introduction

The question of why anything can or should 'mean' anything to anyone seems to belong in the set of fundamental questions that scientific and philosophical pursuits should aim to provide an adequate answer to. Surprisingly, in spite of thousands of years of analytical philosophy and close to 200 years of psychological research on this concept, no sufficiently detailed and prescriptive theory exists that can help us build machines that think. Our view is that artificial intelligence, or at the very least its sub-field of general machine intelligence, calls for this question to be addressed.

Philosophy uses the term *foundational meaning* to refer to the *basic* core concept of meaning [7], as in e.g., "the guitar that your father gave you has

K. R. Thórisson et al. (Eds.): AGI 2024, LNAI 14951, pp. 188–198, 2024.
https://doi.org/10.1007/978-3-031-65572-2_21

'a lot of meaning' to you," and when a loved one says you "mean everything" to them, as well as more basic experiences like breaking out in a cold sweat when being robbed at gunpoint: The *meaning* of such situations *to you* – if you are the one having the experience – is *'foundational.'* This use of the term is different from its use when talking about symbols, signs, and language, which has been called 'semantic' meaning. If a parrot were to squawk at you "I'm gonna *MESS* you up!!", our guess is that you would not be frightened. This is because while you generate semantic meaning for the vocal phrase – which happens to constitute a threat – to you the threat itself has no foundational meaning. The only foundational meaning you generate is of 'a parrot making an empty threat.' A parrot capable of messing up a human is a rather frightening thought; if the situation seems funny, it is due to the sharp contrast between the semantic and the foundational meanings. Generating semantic meaning of e.g. text is a classification task; generating foundational meaning involves much more.

Here we present a computational theory of foundational meaning. The theory already has a rudimentary implementation in a cognitive architecture [9,13].[1] In this paper we describe the theory and detail its foundation and formalisms.

2 Related Work

The study of "meaning" has been a central topic in various disciplines, including philosophy, linguistics, and cognitive sciences. Philosophic study of meaning goes back thousands of years: disquisitions and formulations on the nature of the meaning of words, symbols and ideas can already be found in the works of ancient Greek philosophers such as Socrates, Plato and Aristotle [12].

We agree with philosopher David Lewis [7] that it is important to distinguish Foundational and Semantic theories of meaning. The former refers to why anything should 'mean' anything to someone, while the latter refers to translation of some symbol structure into a pragmatic knowledge structure (e.g. the message "Don't step on the grass"). In our theory they differ by their requirements of information inclusion (the former requires reference to the active goals and plans of the mind that generates the meaning, while the latter may not). We see these two areas of meaning as sharing a large number of features, which we will detail in the next sections.

As mentioned, our theory is about foundational meaning, so any related theory that does not address the question of how meaning is *generated* does not have direct bearing on the particulars of our work described here. This includes the well-known theories of Wittgenstein [17], Grice [6], and many others, whose focus on the meaning of language and symbols renders their theories (mostly) devoid of attempts to outline the particular mechanisms necessary for systems that generate their own meaning. Similar can be said about what have been called Referential theories of meaning [1,2,4,8]. Linguistic theories of meaning concerned with sound-meaning relationships are likewise orthogonally related to

[1] See http://www.openaera.org—*accessed Jun. 1st, 2024.*

efforts of building AI systems that generate meaning. An overview of a wide range of philosophical theories of meaning can be found in [5].

In addressing the concept of intelligence, consciousness, and thought, Dennett proposed the concept of the 'intentional stance' [3], which can be adopted by any third party wanting to explain the behavior of an observed complex system that is goal-directed and is expected to act (or aim to act) rationally. This view is very compatible with our theory; one could say that ours begins where it ends.

Our approach to the subject of meaning is in line with Peirce's Pragmatic Maxim [10,11], which asserts that the meaning of a concept resides in its conceivable practical effects or consequences and is revealed through its potential impact on experience and behaviour.

3 Definitions and Concepts

A **world** W is a formal description of a set of constraints and processes having its own universal clock that establish universal ground truth for an intelligent agent. W consists of a set of variables $V = \{v_1, v_2, ..., v_{||V||}\}$, a set of dynamic functions F, an initial state S_0, and a set of relations $\Re$ between the variables, formally $W = \langle V, F, S_0, \Re \rangle$. The variables represent everything that may change or hold a particular value in the world. The dynamic functions define the *laws of nature* in W and update its *full state*: $S_{t+\delta} = F(S_t)$ for each $t + \delta$ timestep. The dynamic functions consist of a set of transition functions $F = \{f_1, f_2, ..., f_n\}$ where $f_i : S^- \rightarrow S^{'-}$ and $S^-, S^{'-}$ are partial states. *Invariant relations* in W are conditions or properties over the variables of W that remain unchanged as W transitions from one state to another. W is assumed to be nondeterministic, otherwise the concept of 'choice' would be empty (see below). W's clock limits measurements and intervention to a moment in time (the "now" – a demarcated interval), anchoring both to a particular moment (or interval) in time, in relation to other events. Measurements and manipulation of W can only be done in the *now*. No changes can be made to the past; measurements of the past can be made under the assumption of no intervening changes having happened; measurements and changes of the future can be planned through prediction (see below).

A world **state** S is a set of variables in the world, $S \subset W$, that have assigned values (some of which can be measured by an agent through its sensors, and affected through actuators). A **situation** σ refers to the immediate surroundings of an agent, where a subset of a world (a set of states) can be relatively easily measured and affected by the agent at any given time, $\sigma \in W_t$.

An **agent** is an embodied system consisting of sensors and actuators, and a controller (the mind), implemented in some computational substrate (all together this constitutes the agent's *body*). The controller's substrate contains resources (compute power and knowledge) that can be committed to cognitive tasks by an attention process. The body defines the agent's interface to the world, which allows the measuring (perception) of variables in W, through the flow of energy from the body's sensors to the controller, turning it into data, and the flow of energy towards its end-effectors for the execution of atomic actions, by means of commands initiated by the controller for the body's actuators. An agent can be assigned a (set of) top-level drive(s), which define its reason for existence (in the sense that a vacuum cleaner's purpose is to keep the floors clean); all goals and subgoals are derived from this top level (cf. [13]).

A **controller** consists of a set of processes P that can receive an input, $i \in I$, produced by either measuring the world W or by producing reflectable knowledge (inspectable and dissectable thoughts), current state S, at least one goal $g_x \in G$ (implicit or explicit) and output $o \in O$ in the form of atomic actions (selected from a set of atomic possible outputs O), that (in the limit) achieve goal(s) G. We define **autonomy** not as the size of the space of actions that are available to the agent at any point in time, but rather, the ability of an agent to act based only on its own knowledge, without having to 'call home' (for the intervention of an external controller, e.g. a designer or teacher).

An agent $\mathcal{A}$ is **situated** if $\mathcal{A}$ is embodied and positioned in a particular circumstance or situation σ that is subject to limited energy, space and time (LEST). Our theory of meaning pertains only to the physical world (making no claims about abstract or hypothetical ones); a situated agent in our theory is thus always subject to LEST. A situated agent's ***action potential*** is its potential to assert changes on the environment, in the pursuit of its active goals, as limited by the current situation (through LEST).

A **goal** state (positive goal) is a desirable (possibly partially defined) state that the agent could or should reach. Conversely, a failure state (negative goal) is an undesirable state that the agent should avoid. The goals an agent is ultimately expected to pursue are called top-level goals (G_{top}) and derive from $\mathcal{A}$'s drives (see above). A goal can be decomposed into a number of subgoals, which describe and constrain how it can be achieved, resulting in a goal hierarchy with top-level goals at the top and atomic actions at the bottom. At any given time, only a subset of the goal hierarchy is commmitted to being pursued – the **active goals**. A **plan** is a sequence of (atomic) actions extending over a period of time, whose successful execution is expected to update the state of the world according to the agent's goals and subgoals, so that a goal (or subgoal) state is achieved and/or a failure state is avoided. To realize plans, the agent's runtime operation involves generating, evaluating, replacing and committing to goals, while ensuring sensible usage of available resources and knowledge, responsiveness to unexpected events, in a mixed planning/opportunistic manner.

A learning agent's knowledge $\mathcal{K}$ consists of a growing and changing set of goals and **models**, endogenously formed from its experience. Good models allow an agent to systematically predict, affect, explain and re-create phenomena [14]. To do so effectively and efficiently, models must capture **causal relations** between (hypothesized) causes and their effects [15]. The better (effective, efficient, useful) the models are for these purposes, the more **reliable** they are.[2] Causal models describe the evolution of a substate of the world $S \subset W$ as a conditional state

[2] Models of the physical world generated by an embodied agent may always turn out to be incorrect; guarantees of 'truthfulness' cannot be given. Models thus cannot be said to be 'correct' or 'true,' only *useful* and *reliable*.

transition function (the fewer conditions, the more general it is) that applies a (conditional) rule R to S producing the new state S': $S \xrightarrow{R} S'$. Causal models can be combined in various ways, via ampliative reasoning [13], to describe the transformation of one world state S to another S', in e.g. the attainment of goals and whether they are realistically possible or not (one role of intelligence is figuring out which ones are). The application of N causal models, where the input state of each model m is the output state of the previous model $m-1$, forms a *causal chain* of N sub-states potentially reachable by the system. Causal chains are built through a process that applies (non-axiomatic, defeasible) reasoning to models, in the service of goals.

A situated agent's knowledge about a particular situation is **grounded** if an unbroken contextualized chain of causal models can be formed that connects the low-level perceptual data, generated in the 'now' from the agent's situation, to its top-level goals. A chain is "broken" if the reliability of any of the causal models forming the chain is below a minimum threshold; this reliability is explicit metaknowledge that describes the chain itself that can be reasoned over (i.e. supports reflection) like models and goals. To consistently pursue and achieve goals, predict and model its situation, $\mathcal{A}$ must have some minimal set of causal and relational models of the tuple $\langle \mathcal{A}, \sigma \rangle$. The agent's model of self is built in the same way as other knowledge, consisting of models of its own mind and body, initially derived from its seed (the knowledge the agent was born with) and progressively refined, expanded and generalized through experience.

Prediction is an ongoing cognitive process Pr that produces hypothetical future states S that can be used as the basis for generating new goals G and plans Pl. Pr implements (non-axiomatic, defeasible) abduction and (non-axiomatic, defeasible) deduction processes, both of which rely on models of cause-effect relations. Due to physical limitations in cognitive resources, an agent must select which predictions to make for any period of time.

Uncertainty exists in all knowledge, stemming from imprecision in *measurement* (because taking reliable measurements takes time, and time is always limited), *predictions* (because they are produced from defeasible knowledge), and *control* (due to unknown factors – 'noise'). A **choice** is made by an agent through a commitment of resources to one option from a set of mutually exclusive alternatives. This includes both thinking and interacting with the world.

4 Meaning Generation

Now we can outline how foundational meaning is generated in an autonomous situated agent. The following applies equally to agents found in nature and those manufactured in a lab.

> ***Definition.*** The *foundational meaning* of a datum $\mathcal{I}$ to agent $\mathcal{A}$ in situation σ is constituted by a causal coupling of $\mathcal{A}$'s (model of its) *situated future* to (its models of) its own action potential – what it considers itself to be capable and not capable of doing, in light of $\mathcal{I}$, in the form of (active and non-active) goals and plans, represented in an explicit hierarchy of relevant knowledge $\mathcal{K}$ – such that, based on $\mathcal{A}$'s active goals $\mathcal{G}$ and resulting new goals G and plans Pl, any relations of $\mathcal{I}$'s to $\mathcal{A}$'s knowledge can be (causally) traced to $\mathcal{A}$'s top-level goal(s), G_{top}.

By 'datum' we mean an information structure containing parts that are recognizable (at least to some minimal extent) by an agent. The reason it must have recognizable (classifiable) parts is that without any such features, the datum cannot be processed by the agent's cognition, and would thus be meaningless ($\mathcal{I}$ would be 'incogitable' – invisible to the agent's cognitive mechanisms).

Causal knowledge is a necessary requirement for systematic, efficient and effective control of a phenomenon. An agent $\mathcal{A}$'s *situated future* consists of its *predictions of what may and may not happen in its spatio-temporal proximity*, given its knowledge of cause-effect relations relevant to σ. The causal coupling centers around causal relations, and related relevant knowledge, that form an unbroken (non-axiomatic, defeasible) chain that connects top-level goals and low-level perceptions, contextualizing $\mathcal{I}$ in $\mathcal{A}$'s knowledge network. The more reliable the weakest causal link is in this chain, the *stronger is the grounding* of the meaning of $\mathcal{I}$ to $\mathcal{A}$. $\mathcal{A}$'s *action potential* is the set of actions (including measurements) that can be performed by $\mathcal{A}$ with respect to $\mathcal{I}$ in a given situation σ (σ defines which variables are observable and manipulable by $\mathcal{A}$ and σ thus constrains $\mathcal{A}$'s action potential). We refer to this overall information content as an "$\mathcal{M}$-structure."

Creating an $\mathcal{M}$-structure involves reasoning (non-axiomatic, defeasible deduction, abduction, induction and analogy – in a variety of combinations). Such reasoning also determines whether the chain's strength, whether it is unbroken, and the whether, and how, certain causal models of the world subsume others. Reasoning is also necessary to figure out implications of $\mathcal{I}$ in the present situation σ, which unavoidably consists of models of relations between diverse relevant components – sensors, actuators, objects in the environment, forces, obstacles, etc. – forming a hierarchy of goals, models, plans, parts and wholes.[3] The requirement for manipulable models of whole-part relations and constraints means that the knowledge must be (partly or fully) symbolic;[4] the existence of

[3] For instance, a shoe is made up by parts like laces, soles, etc.; causal relations define what can and cannot be done with the shoe and its parts in particular situations, given certain constraints (e.g. whether we are wearing them or just looking at them).

[4] For practical reasons, including memory storage and compute power, there will always exists a level of detail below which relevant information will not be strictly symbolic, and another lower one below which it will be completely sub-symbolic. This is most obvious for low-level perceptual data, e.g. vision, which may involve bandwidths of 2 megabits per second or more.

novelty and uncertainty in a non-axiomatic world means that the process and runtime of the reasoning cannot be known beforehand, so it must be ampliative and learnable.[5] Learning to do ampliative reasoning, in turn, cannot be done without reflection over both the domain knowledge and the reasoning mechanisms themselves, which calls for a transparent compositional explicit representational scheme.

To ensure a certain level of alertness and action capacity, a situated agent must generate meaning continuously. Meaning generation is a situated thought process that is linked to the 'now' via active goals $\mathcal{G}_t$, where t is a (short) interval. In other words, grounding – grounded cognition – happens via linking a situation σ_t to active goals $\mathcal{G}_t$ using (reliable) causal models.

The **foundational meaning** $\mathcal{M}$ of $\mathcal{I}$ for agent $\mathcal{A}$ in situation σ is thus captured by the $\mathcal{M}$-structure:

$$\mathcal{M}^{\mathcal{A}}_{now}(\mathcal{I}) = Pr_t(\mathcal{I}, \sigma_t, \mathcal{K}) \rightarrow G_{t'} \rightarrow Pl_{t''} \quad (1)$$

where: *now*: *minimum time interval for meaning generation;* $now = t'' - t$
$\mathcal{I}$: *information structure* $\mathcal{A}$: *agent (embodied controller)*
Pr: *predictions* σ: *situation as modeled by the agent*
$\mathcal{K}$: *the agent's prior knowledge deemed relevant to* $\mathcal{I}$ *and* σ
$G_{t'}$: *new or updated goals (active and/or passive) created from* Pr_t
Pl: *new plans created from* $G_{t'}$

'Now' spans the time from the measurement when $\mathcal{I}_t$ is received, to the time that a plan $Pl_{t''}$ has been produced from the goals $G_{t'}$, that in turn derive from the predictions Pr_t.

Question: *Why are predictions part of the* $\mathcal{M}$*-structure?*

An intelligent agent will always be in pursuit of a set of goals, in the very least due to the constantly changing world it's situated in. The resources needed to achieve goals, including time, energy, space, and knowledge, the situation's 'now' puts constraints on their achievement. An implicit goal of an intelligent agent is keeping its action potential high (maximizing its potential choices). When a controller is presented with a new piece of information, $\mathcal{I}$, the relation of this information to its active goals must be assessed, to see how the action potential for them may be impacted. This impact is always in the future; predictions and plans are a way to detail the shape of such impact. Pursuit of impossible goals is a waste of a controller's resources, so it must be capable of making reasonable predictions about which goals are possible and more sensible than others, and why. These predictions must say something about the future state of the situation σ and the agent itself – including its knowledge, embodiment, and presently active goals $\mathcal{G}$. If an important active goal, e.g. staying alive, is predicted to be heavily impacted or categorically prevented, this has profound implications for the agent's existence (including the fact that all of its other

[5] In our approach, ampliative reasoning' includes (non-axiomatic) deduction, abduction, induction and analogy. These must be dynamically chosen at runtime based on situation and active goals.

active goals will also be prevented). Meaning in a thinking system is thus carried by predictions.

▷ *Predictions participate in the $\mathcal{M}$-structure by ensuring the reliability of committed goals and plans.*

Q: *How does an agent's perception of a situation affect meaning generation?*

The current situation, including an agent's embodiment, constrains what the agent can sense and affect at any particular point in time – and this is critical for determining what knowledge to use for producing goals and plans. Creating useful (new) models is also dependent on perception and classification of the present context, and models are the smallest relevant unit for generating meaning, supporting prediction and action. Predictions that are not grounded in the 'now' and don't take into account recent changes in the world will be based on outdated information, resulting in incorrect predictions, which in turn will produce invalid goals and plans.

▷ *The agent's model of the current situation is part of the $\mathcal{M}$-structure.*

Q: *How does an agent's prior knowledge participate in determining meaning?*

To make sense of new information $\mathcal{I}$, an agent has to relate it to its knowledge, including its active goals. This includes dissecting $\mathcal{I}$ into its constituent components and classifying the present situation σ. The only way to do either is by using existing knowledge. An agent's top-level goals incarnate its mission; any active subgoals will by definition be related to this mission. Information that may help or hinder it in pursuing its active goals will thus be relevant to its existence. For example, if someone points out a pedestrian to an agent driving a car, extracting their walking direction may predict them to be on a direct collision course with the car. The classification in this example requires knowledge about the look and behavior of pedestrians, their mode of locomotion and speed, and the same for the controlled vehicle. Avoiding harm to others may be one of the driver's negative goals. The meaning generated from this dissection – the potential for causing future harm – thus xaffects an important top-level goal of the agent. There is no condition under which such classification could be done without some knowledge about the phenomena in question, at various levels of detail, in light of the goal.

▷ *The agent's prior knowledge is part of the $\mathcal{M}$-structure.*

Q: *Why are new/updated goals part of meaning generation?*

Full autonomy requires a system to generate goals and subgoals on its own. At any moment in time, only a subset of the system's goals are active; active goals can be seen as a set of (multi-dimensional) attractors (consisting of subgoals) that steer the agent's behavior, with the intent of transforming a situation to align with its goals. The autonomous goal-generation process is driven by predictions and reasoning over models of the current knowledge of the situation. Any perturbation of a path towards a goal may require changing some subset of a plan or re-planning from scratch. Predicted perturbations may also require a goal to check at a later time whether the prediction came true. In either case, the end-product of such efforts would be described by a new or updated goal.

The meaning is carried not only by the predictions of future states of the present situation but also the relation of these predictions to the presently active goals and, ultimately, their relation to the agent's top-level goals. Hence, updated and new goals are a necessary step in the meaning generation.

▷ *Updated/new goals are part of the $\mathcal{M}$-structure to ensure an agent's mission.*

Q: *Why are plans part of meaning generation?*

Any situation's meaning – to the system itself – is captured by the system's predictions of how the situation develops into the future, and how this development affects the currently active goals. Given the predicted effect of the present situation σ_t on the active goals $\mathcal{G}_t$, and adjusted goals $G_{t'}$ that take those predictions into account, new plans Pl_t provide actionable descriptions of how these could be achieved. The generation of plans, and their form, is a necessary part of the meaning generation process because it outlines what is possible and what is impossible under the constraints presented by σ and $\mathcal{I}$. Plans thus provide a mechanism for distinguishing between useful and useless (meaning*ful* and meaning*less*) information with respect to active goals and the situation, in light of available knowledge and resources. This, then, is the "end of the line" of the meaning generation process, feeding back to new predictions, which then produces new updates to the present meaning structure.

▷ *Plans are part of the $\mathcal{M}$-structure by clarifying the importance of information.*

Q: *Why is meaning always bound in the 'now'?*

Because the physical world has an immutable 'ticking clock.' Thinking is control, and control is about making choices in light of options, under the constraints of the world clock. The past cannot be changed, and only choices made in the 'now' can affect the future – all agents in the physical world are prisoners of the 'now.' Actions are thus only to be enacted in the 'now.' Simply by 'ticking,' the world clock changes the action potential of all agents, independently of what they do. This means that the foundational meaning of an agent's knowledge will change eventually. Meaning generation is an implemented (physical, "cyber-physical") computational process. Computation cannot happen in the past or in the future: for an autonomous controller the 'now' thus serves as the anchor point for interpreting the current situation; in the 'now,' fixated at a particular point in time, because for meaning to exist, the outlined computations must be performed. In other words, the generation of meaning is a dynamic process that has to be executed by a concrete (physical) computing agent. Regardless of previous events, meaning depends only on the information and knowledge currently available to an agent. Semantic meaning thus depends on foundational meaning.

▷ *Semantic and foundational meaning are always anchored in the 'now.'*

Q: *Is there a difference between meaning and autonomy?*

What sets intelligence apart from all other phenomena, and thus defines it, is its ability to handle novelty [16]. Meaning is the causal linking of novelty, knowledge, a situation in the now, predictions, goals and plans. Meaning generation involves the effective and efficient outcome of this process – we say that

an agent that can do this reliably and repeatedly is able to extract the meaning of situations. This meaning extraction capability allows an agent, in turn, to act autonomously. Meaning generation is not needed if autonomy is not desired or needed; without autonomy there is no need to generate meaning. Autonomy and meaning are in fact two sides of the same coin: *Autonomy without meaning generation is meaningless; meaning generation without autonomy is pointless.*
$\triangleright$ *Autonomy* $\equiv$ *Meaning.*

5 Conclusions and Future Work

Meaning and meaning generation are key aspects of highly autonomous, generally intelligent systems. This work contributes to bridging a gap, for too long left open, between theories of meaning and the design of systems that can generate and handle meaning. Our theory identifies precisely the qualities necessary for a system to generate meaning for itself; the hope is to pave the way for a new class of intelligent systems showing unprecedented autonomy and generality. A more exhaustive treatment of the implications of this theory, restricted here for reasons of space, is deferred to future work.

Acknowledgment. The authors would like to thank the GMI team at Reykjavik University, Bridget Burger and the anonymous reviewers for insights and discussions.

Disclosure of Interests. The authors have no competing interests to declare that are relevant to the content of this paper.

References

1. Alston, W.P.: Meaning. In: Edwards, P. (ed.) The Encyclopedia of Philosophy, pp. 5–233. Macmillan (1967)
2. Braun, D.: Russellianism and explanation. Philos. Perspect. **15**, 253–289 (2001)
3. Dennett, D.: Brainstorms. M.I.T. Press, Cambridge, MA (1978)
4. Frege, G.: Sense and reference. Philos. Rev. **57**(3), 209–230 (1948)
5. Gelepithis, P.: Survey of theories of meaning. Cogn. Syst., 141–162 (1988)
6. Grice, H.: Studies in the Way of Words. Harvard University Press (1989)
7. Lewis, D.K.: General semantics. Synthese **22**(1–2), 18–67 (1970)
8. Lyons, J.: Language, Meaning, and Context. Fontana, [London] (1981)
9. Nivel, E., Thórisson, K.R., Dindo, H., Pezzulo, G., et al.: Autocatalytic endogenous reflective architecture. Technical RUTR-SCS13002, Reykjavik University School of Computer Science, Reykjavik, Iceland (2013)
10. Peirce, C.S.: Pragmatism as a Principle and Method of Right Thinking: The 1903 Harvard Lectures on Pragmatism. State University of New York Press (1997). Turrisi, P.A.: (ed.)
11. Peirce, C.S., de Waal, C.: Illustrations of the Logic of Science. Open Court, Chicago, Illinois (2014)

12. Prior, A.N.: Correspondence theory of truth. In: Edwards, P. (ed.) The Encyclopedia of Philosophy, vol. 2, pp. 223–224. Macmillan (1967)
13. Thórisson, K.R.: Seed-programmed autonomous general learning. Proc. Mach. Learn. Res. **131**, 32–70 (2020)
14. Thórisson, K.R., Kremelberg, D., Steunebrink, B.R., Nivel, E.: About understanding. In: Steunebrink, B., Wang, P., Goertzel, B. (eds.) AGI -2016. LNCS (LNAI), vol. 9782, pp. 106–117. Springer, Cham (2016). https://doi.org/10.1007/978-3-319-41649-6_11
15. Thórisson, K.R., Talbot, A.: Cumulative learning with causal-relational models. In: Iklé, M., Franz, A., Rzepka, R., Goertzel, B. (eds.) AGI 2018. LNCS (LNAI), vol. 10999, pp. 227–237. Springer, Cham (2018). https://doi.org/10.1007/978-3-319-97676-1_22
16. Wang, P.: On defining artificial intelligence. J. Artif. General Intell. **10**, 1–37 (2019)
17. Wittgenstein, L.: Philosophical Investigations. Basil Blackwell, Oxford (1953)

Causal Inference in NARS

Bowen Xu and Pei Wang(✉)

Department of Computer and Information Sciences, Temple University, Philadelphia, USA
{bowen.xu,pei.wang}@temple.edu

Abstract. Humans engage in causal inference almost every day, however, the term 'causation' is still quite ambiguous, and few AI systems provide a comprehensive and satisfactory solution to causal inference. In this paper, we adopt the primary meaning of causation, *i.e.*, *prediction*, and argue that in different contexts other demands are attached to it. We describe the approach of causal inference in NARS and present some working examples, both at the sensorimotor and abstract levels. The theoretical and practical consequences are quite different from traditional AI approaches.

Keywords: Causal Inference · Non-Axiomatic Reasoning System · Prediction · Explanation

1 Introduction

Causal inference has attracted wide interest over the years in Artificial Intelligence (AI), though it has a rather long history in philosophy, psychology, economics, physics, and so on. On the one hand, causal inference seems to be a fundamental capability of intelligent agents to interact with the world; on the other hand, there is no consensus on what causation is, let alone the mechanism of causal inference [12]. Some people consider causation a fat concept with diverse meanings in different contexts [2,3]. Nevertheless, over the years, some AI researchers (*e.g.* [10]) believed that causal inference would bring about a scientific revolution, attracting broad interest in enabling AI to make causal inferences.

Traditionally, causation is viewed as an objective relation between events, and various definitions were proposed from distinct perspectives (*e.g.*, regularity, counterfactualism, interventionism, *etc.*) [2,7,12]. In contrast, causation has been treated as a subjective relation in Non-Axiomatic Reasoning System (NARS) [6, 14] where no assumption is made on the "true cause(s)" of an event [18], though predicting the future or explaining the past based on the current situation is referred to as "causal inference".

In this paper, we further discuss the nature of causal inference in NARS. We start by briefly reviewing previous works, especially in AI, and then proposing our positions and perspectives on causation and causal inference. Finally, some working examples are provided to show that NARS can do causal inference in a unified manner: both in sensorimotor procedure and abstract-level reasoning.

K. R. Thórisson et al. (Eds.): AGI 2024, LNAI 14951, pp. 199–209, 2024.
https://doi.org/10.1007/978-3-031-65572-2_22

2 Previous Works

Causation and causal inference have been discussed in many domains. For example, the representative of the subjective view of causation is David Hume, a philosopher who believes that it is impossible to deduce consistent laws from existing observations and that 'causation' is nothing but a human mental habit or even an illusion. In contrast, the representative of the objective view of causation is Pierre-Simon Laplace, who saw the present state of the universe as the result of its previous state and the cause of its subsequent state so that an omniscient and omnipotent intellect could accurately predict the future according to causal laws.

In Artificial Intelligence, the most influential approaches to causal inference are probabilistic, as exemplified by the *structural causal models* (SCM) proposed by Judea Pearl [9], where causes and effects are nodes in a graph, and directed connections represent causal relations. In Pearl's theory, a "causal model" is composed of a set of "causal assumptions" (or prior knowledge) provided by human scientists, and causal inference aims to evaluate the extent of the truthfulness of an implication. Pearl formalizes *intervention* using "*do*-calculus" [10], by which a conditional probability is adjusted to eliminate cognitive bias. However, an often-overlooked implicit assumption in Pearl's theory is that future observation will follow the same regularity as the past. With this assumption, given a causal model and collected data, an actual estimate of effects can be obtained as if it did some interventions to collect more data. As a result, one is able either to know "true effects" by mere observation or to know such estimation is unavailable. The same assumption is made in Pearl's theory on counterfactual.

Instead of assuming an available causal model, some researchers focus on discovering graph structures as possible causal explanations for data, and this ramification of research is called "causal discovery" [13,19]. Solving a causal discovery problem was defined as "*recovering the true graph from the given data set*" [19], implicitly assuming the existence and availability of a *true* causal graph. In some cases, the aim of causal discovery is "*revealing causal information by analyzing purely observational data*" [4].

We argue that the traditional views on causal inference mentioned above do not apply to *AGI* under our conception [16]. It does not mean that their works are wrong, but that, as far as *causal inference* is concerned, AGI systems should work under different conditions from the previous ones. Firstly, AGI systems actively acquire data rather than passively observing, as opposed to causal inference [10] in the traditional sense. Secondly, AGI systems learn causal relations among events by themselves without given knowledge. Thirdly, AGI systems may face situations where future observations are not guaranteed to conform to past findings, so that "true (causal) graphs" [19] should not be assumed in causal discovery. Finally, there cannot be a causal model containing all possible events, as many events are known to the system only after their occurrence.

While in many traditional models "causal inference" only covers *causal estimation*, in NARS that phrase also covers *causal discovery*, as the two functions are carried out by the same mechanism, rather than handled separately as in the traditional models.

3 The NARS Approach

Because NARS is based on the *Assumption of Insufficient Knowledge and Resources*, the objective view of causation cannot be accepted [14]. There is no practical difference between saying "Objective causal laws do exist, but we can never get them" and saying "There are no objective laws". It is even impossible to say "Our knowledge is an approximation of objective laws" because such a statement makes sense only if the degree of approximation can be determined. The historical success of science does not mean that we can obtain concrete "objective causation", because no theory, however successful in the past, can logically guarantee the correctness of all its predictions for the future, as Hume argued [8]. But we are also against *agnosticism* because the result is often the abandonment of all effort and even the denial of the value of thinking. The key here is how to interpret the meaning of *knowing*. We believe the *knowledge* of a cognitive subject depends on the cognitive ability and mechanism of the subject, so it is improper to take "knowable" as meaning "be able to be obtained in a complete, accurate, and irrefutable description" of an object. All objects are knowable, but all knowledge is subject to revision or rejection.

In the case of causal knowledge, admitting that they are "mental habits" does not necessarily make them less valuable. The adaptability of intelligent systems requires effectively predicting the future and controlling it as much as possible. This requires finding regularities in experience, even though the results are not guaranteed to hold in the future, or they have known counter-evidence. The rationality in this situation is to be *as good as possible*, rather than *absolutely good.* "Causality" under this interpretation is not the objective connection between events (even what counts as an "event" and whether there is a "connection" depends on the cognitive subject), but the rational connection between events that the cognitive subject can use to make predictions. To try to find a "cause" for an event is to explore the possibility of predicting and controlling it. This is a basic requirement for adapting to the environment, and an intelligent system must strive to do so, no matter how well it can do it under the circumstances.

Based on this perspective, "causation" is not a *logical* relationship in NARS, but is usually represented as acquired implicational statements with temporal order and other domain-specific attributes. In the simplest case, the statement "A is the cause of B" can be represented as "$A \not\Rightarrow B$", where "$\Rightarrow$" is implication, and "/" indicates that A occurs before B. Similar to other types of knowledge in NARS, such causal knowledge can be acquired either directly from external input or derived through the system's reasoning activity.

One common way of deriving causation is "temporal induction", where the conclusion "A always leads to B when A happens before B" is generated from observed instances of "A happens, and then B happens". Like other inductive conclusions, the initial *confidence* of such a conclusion is relatively low, which can be referred to as a "hypothesis". Only when similar observations are repeated numerous times, a reliable conclusion can be obtained. In contrast, the deductive relationship "$A \not\Rightarrow B$", derived from "$A \not\Rightarrow C$" and "$C \not\Rightarrow B$", has a higher *confidence* value.

The reason for causation to be represented in this form is that the original meaning of "$A \not\Rightarrow B$" is "prediction", that is, to infer "B is about to happen" from "A has happened", and this is the major role of causation in adaptation. Therefore, in NARS, there is no separate "causal reasoning" mechanism, but the temporal order of events is taken into account in the inference rules dealing with implication. This is because what we commonly call "causation" does not have a definite logical meaning as "$A \not\Rightarrow B$", but largely depends on contextual factors such as domain and purpose.

Based on this view, let's analyze the given options and determine which one is most suitable to be considered "the cause" of the collective increase in stock prices in a certain sector of a stock market –

1. increased demand for related products in the recent period,
2. technological breakthroughs suggesting the imminent release of new products,
3. the release of statistical data earlier that day enhanced investors' confidence, and
4. a large number of people buying stocks in that sector on that day.

Journalists may choose (3) in their coverage, while economists are more likely to choose (1) or (2) and start debating which factor is more important. They will all think that (4) is self-evident and therefore not worth mentioning. Since the logic behind these choices is "to infer that B is about to happen from the fact that A has already happened", it is suggested that "A is the cause of B". Besides the necessary factor of "prediction", there are several other relevant factors, although they are not taken into account in every context:

- **Necessity:** In some fields, A is a sufficient condition for B (*i.e.*, "$A \not\Rightarrow B$"), while in some other fields, A is a necessary and sufficient condition for B (*i.e.*, "$A \not\Leftrightarrow B$").
- **Reality:** In some fields, causation is meaningful only when both A and B have occurred, while in others, they can be hypothetical or imagined events (*e.g.*, "What caused me to get wet was that I did not have an umbrella").
- **Counterexamples:** In some fields, a single counterexample can reject a proposed causal relation, while in some other fields, a causal relationship can have exceptions.
- **Controllability:** Some contexts require that A must be an action of an agent or under its control to be considered the cause of B.
- **Reproducibility:** In certain domains, causal relations can only be established between repeatable events, rather than between one-time events.
- **Explanation:** In some fields, a correlation between A and B alone is not enough to establish a causal relation between them, and an explanation is required (*e.g.*, A causes M and M causes B).

- **Temporal Distance:** The temporal distance between A and B may affect whether A is considered the cause of B, otherwise, the "Big Bang" could have been the cause of everything.
- **Exclusivity:** In some contexts, there might be only one cause for an event, while in others, multiple causes may be accepted.

The requirements above are all reasonable, but none of them is universally required when "causation" is considered. It would be nice to meet all the requirements, but if there is no such knowledge that meets them all, "no prediction at all" could be worse than wrong predictions. Thus, NARS allows for the coexistence of different types of causation in various contexts, each of which can attach additional requirements besides "supporting prediction". As a result, the specific meaning of 'causation' needs to be learned in each domain (physics, medicine, geology, economics, law, *etc.*), which may have domain-specific conditions. For example, "Gravity is the cause of a ball falling" can be expressed in NARS as "$(gravity \times falling\text{-}ball) \rightarrow physical\text{-}cause$", where "*physical-cause*" is an acquired relation that represents "cause" in physics. In our opinion, trying to deal with causation with multiple interpretations in a single manner is the main cause of the confusion in current research.

According to Piaget's theory, the concept of causation in infants begins to form through their understanding of the consequences of their actions [11]. In NARS, it can be represented as "$(S_1, \Uparrow Op) \not\Rightarrow S_2$", which means "After seeing S_1, if I perform operation $\Uparrow Op$, I will see S_2". Through analogy and inductive reasoning, it can be generalized to "$(S_1, S_3) \not\Rightarrow S_2$", which means "After seeing S_1, if any person performs the operation S_3 (which corresponds to my $\Uparrow Op$), the person will see S_2", where "the operation corresponding to my $\Uparrow Op$" is considered as an event S_3 (since it is not performed by me). This relationship can be further generalized to "$S_4 \not\Rightarrow S_2$", where S_4 is the name of the compound event '(S_1, S_3). In this form, there is no need for any cognitive subject to act as the cause of a result. Although a mature cognitive system may acquire causal knowledge directly from communicating with other systems, without going through the process of "from self to other agents then to events", this explanation can still apply to many phenomena related to causal inference. Note that the procedure of causal inference above also involves self-awareness [1], but due to the word limit, self-awareness is not discussed in this paper (refer to [17] for further information).

A direct consequence of this view on causation is that no intelligent system can accurately foresee all the consequences of an event, including its actions. As the saying goes, "What seems like a loss may be a gain." However, this is not a reason to refrain from predicting or taking action, and all decisions should be based on weighing the pros and cons of foreseeable consequences.

4 Working Examples

In the following, several working examples of causal inference in NARS are described and analyzed.

4.1 Sensorimotor Integration

An example is related to the sensorimotor procedure, aiming to play the Cart-Pole game in reinforcement learning. An agent should learn some general rules, such as that if the pole leans to the right, the cart will move to the right. An implementation of NARS [5] is used to play the game. The performance of the system is shown in Fig. 1, where the score indicates the proportion of time the system successfully balanced the pole over the past 100 moments. By checking its memory, the following beliefs used in decision-making are acquired by the system

$$((angle1, \Uparrow left) \not\Rightarrow good).$$
$$((angle2, \Uparrow right) \not\Rightarrow good).$$

These are causal beliefs about the environment, stating under what condition, by taking what operation, what effect would be caused. Based on the causal beliefs, NARS learns to perform satisfactorily in this environment.

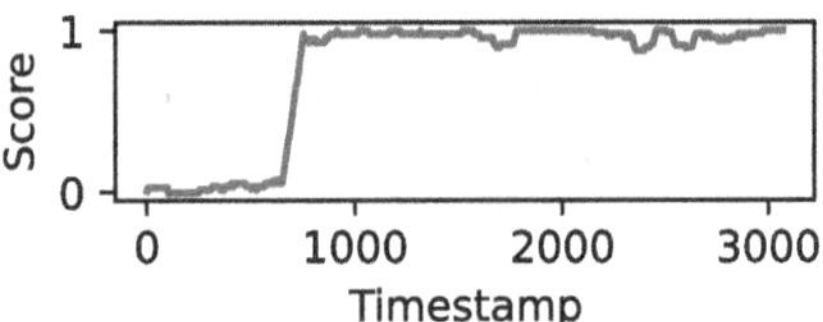

Fig. 1. The performance of NARS playing the CartPole game.

4.2 Simpson's Paradox

Besides the sensorimotor procedure, NARS can do causal inference in a more abstract level, following the same logic. An example of intervention analyzed by [10] is Simpson's paradox. However, in NARS, quite different conclusions are drawn from Pearl's theory of causal inference. For example (see Table 1), the rates of heart attacks with drugs are 7.5% for women and 40% for men, and the rates without drugs are 5.3% for women and 30% for men. However, the rates of heart attacks for all are 18.3% with the drug and 21.7% without the drug, that is, the drug increases the risk of heart attack for men, and it also increases the risk for women, but it decreases the risk for the population as a whole!

In [10], Pearl explained that this is mainly because gender is a "confounder", and a person with different genders may have different selections of whether to take the drug, and the risk of heart attack is also directly influenced by gender. They suggested to do statistical adjustment by using the back-door criterion, as a result, $P(H|do(D)) = 0.238$, and $P(H|do(\neg D)) = 0.177$. In this way, they said the correct conclusion can be drawn that the drug still harms the whole population. Despite the number game behind the paradox, let us check the implicit

Table 1. Fictitious data illustrating Simpson's paradox. This table stems from [10] (Table 6.4).

	Control Group (No Drug)		Treatment Group (Took Drug)	
	Heart Attack	No Heart Attack	Heart Attack	No Heart Attack
Female	1	19	3	37
Male	12	28	8	12
Total	13	47	11	49

assumptions behind this statistical adjustment. The interpretation of *do*-calculus in Pearl's causal inference theory is that a certain intervention is executed without really doing it so that one can evaluate the "true causal effect" on observed data without further experiments. It is assumed that the data are perfect and sufficient so that regularity would hold if more observations were acquired by action.

In NARS, we have a different explanation and argue that the two conflict beliefs are reasonable to occur in the human mind. Given the two premises

$$((male \wedge D) \not\Rightarrow H).\langle 0.4; 0.95\rangle$$
$$((female \wedge D) \not\Rightarrow H).\langle 0.08; 0.98\rangle$$

and the compositional rule[1], it is derived that

$$(((male \vee female) \wedge D) \not\Rightarrow H).\langle 0.03; 0.93\rangle$$

Similarly, we have "$(((male \vee female) \wedge \neg D) \not\Rightarrow H).\langle 0.02; 0.93\rangle$". It shows that the frequency of people who took the drug and then suffered heart attacks is smaller than that of people who did not take the drug, *i.e.*, the drug increases the risk of heart attacks. In contrast, from direct observations,

$$(D \not\Rightarrow H).\langle 0.18; 0.98\rangle$$
$$(\neg D \not\Rightarrow H).\langle 0.28; 0.98\rangle$$

taking the drug seems better. The two contradictory beliefs coexist in NARS' mind, thus it is a paradox from its view.

4.3 Intervention and Counterfactual

To better illustrate causal inference in NARS, the example shown in Fig. 2 can be used. Suppose that the relevant events mentioned in the following have been obtained from perception: a remote control, C, can be manipulated by a system (like NARS, or a human being) who wants to watch TV shows. Once the system presses a button on C, the corresponding event (as an operation) $\Uparrow P$ occurs in

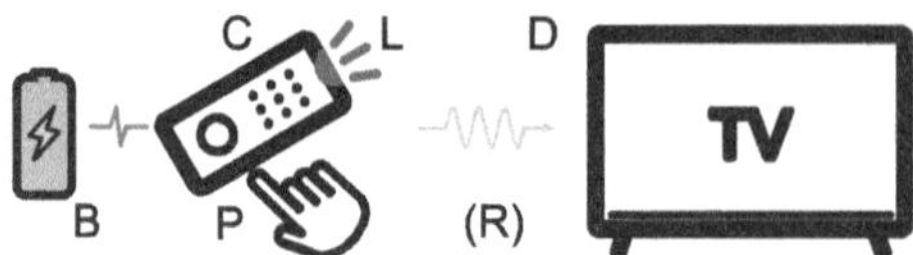

Fig. 2. An Example of Causal Inference

its mind. Subsequently, the indicator light flashes once (event L), and then the TV is turned on and displays a picture (event D).

The system constructs the knowledge "$(\Uparrow P \not\Rightarrow D)$" through *temporal induction*: given events $\Uparrow P$ and D as well as their occurrence time, "$(\Uparrow P \not\Rightarrow D)$" is implied. The *confidence* value is low at first (as it has been tested only once), so it can be viewed as a hypothesis, and the *confidence* value gets higher as more evidence is observed. In most cases, this knowledge is more than enough for the system to predict the future and explain the past, though similarly, it can also form an "explanation chain" for "$\Uparrow P$ causes D": it holds because "$\Uparrow P$ causes L" and "L causes D", that is, "$(\Uparrow P \not\Rightarrow L)$" and "$((\Uparrow P, L) \not\Rightarrow D)$".

The system usually considers $\Uparrow P$ as the cause of L, rather than "$(B, \Uparrow P)$" (where B indicates the existence of a battery in C) because the former is simpler than the latter and the battery is usually there; however, if $\Uparrow P$ by is not followed by L as the battery is missing, "$(B, \Uparrow P)$" is better to be viewed as the cause, because it is more correct. Similarly, when $\Uparrow P$ occurs but D does not subsequently occur, "$(\Uparrow P, L)$" is better to be explained as the cause of D.

There is no "true cause" of TV's displaying D, but there are satisfactory explanations for it. Infants may never generate explanations "Pressing the button $\Uparrow P$ causes an infrared ray R" and "The infrared ray R causes displaying D"; similarly, ill-educated adults may not know "$\Uparrow P$ causes electric current", "the current causes electronic transitions in an infrared emitting diodes", and "electronic transitions produce infrared ray R"... There can be an extremely long chain for explanation. Nevertheless, in daily life, satisfactory explanations are enough to be viewed as causal relations.

Intervention: *Intervention* is nothing but *operation* in NARS. In the context of causal inference, *operation* as *intervention* plays two roles – actively acquiring new knowledge, and filtering competing hypotheses in decision-making. By *operations*, the system changes the environment to construct new knowledge or to collect more evidence for existing knowledge, and some knowledge might not be easy to obtain by pure observation. In this example, the system cannot know "$((\Uparrow P, L) \not\Rightarrow D)$" until actually executing the operation $\Uparrow P$.

Counterfactual: Through exploration, the system finds that "battery" is necessary for L, *i.e.*, "$(B, \Uparrow P) \not\Rightarrow L$"; the battery can be installed through the action

[1] $\{T_1 \Rightarrow M, T_2 \Rightarrow M\} \vdash (T_1 \vee T_2) \Rightarrow M\langle F_{int}\rangle$ [15]. It is isomorphic in temporal inference.

"$\Uparrow$*install*", *i.e.*, "($\Uparrow$*install* $\not\Rightarrow B$)". However, when the system presses the button (*i.e.*, $\Uparrow P$) and the light does not work as anticipated (*i.e.*, $\neg L$), the system might observe the lack of battery (*i.e.*, $\neg B$). Since "if B were true, as a *counterfactual*, then L would also be true", and installing the battery (*i.e.*, $\Uparrow$*install*) implies B, it is derived that if the system installed the battery, the light would flash, that is, "(($\Uparrow$*install*, $\Uparrow P$) $\not\Rightarrow L$)". This is a preliminary form of *counterfactual* reasoning in NARS, although the related inference rules have existed in the logic [15].

Counterfactual reasoning exactly means reasoning with a statement that is contrary to the fact. In this example, $\neg B$ is the *fact* (according to its *truth value*), while "$(B, \Uparrow P) \not\Rightarrow L$" is a basis for concluding of "what if B".

More generally, suppose the system has a belief "($\Uparrow$*assume*(B) $\not\Rightarrow B$)", and the system presses the button, then an implication is "($\Uparrow$*assume*(B) $\not\Rightarrow L$)", and other conclusions can be drawn with the precondition "$\Uparrow$*assume*(B)", such as "($\Uparrow$*assume*(B) $\not\Rightarrow D$)". It means that, even if event D does not actually occur, the system knows "it would have occurred if the remote had the battery".

5 Conclusion

In this paper, we focus on the issue of causal inference in NARS. We argue that the concept "causation" has different meanings in different contexts or disciplines, while the intersection among them is "prediction". Additional demands are attached to "prediction" in various contexts.

In NARS, a *cause* is a kind of *explanation* (namely, *causal explanation*) of an *event*. *Causal explanation* is one of various types of *explanations*, while a *cause* is selected among multiple premises of an *event* (as the conclusion) according to different contexts. The *event* as the conclusion here is also named *effect*. *Causal relation* is the relation between a *cause* and a corresponding *effect*, however, it is not a built-in logical relation, since it has quite different meanings under different contexts and domains. Nevertheless, the original form of causal relation is prediction, and in a specific context, it is a type of *acquired relation*.

We present some working examples of causal inference in NARS, though they do not involve inferring with *acquired relations*. We show that causal inference happens at both the sensorimotor level and the abstract level: the system learns some predictive implications, which can be viewed as *conditioned reflex*, from sensory signals; we address the issue of Simpson's paradox and argue that the cognitive bias also occurs in NARS, and this is reasonable under the *Assumption of Insufficient Knowledge and Resources*; we design a circumstance showing intervention and counterfactual reasoning happen in NARS. By intervention, the system actively obtained new evidence from the environment, so that new knowledge is acquired, which cannot be done by pure observation. Intervention also helps to filter competing hypotheses in decision-making, but we do not discuss this aspect in detail for lack of space. Counterfactual reasoning is about reasoning on the statements that are *false* actually, deriving possible consequences as if those were *true*.

The philosophical positions on causation and the practice of causal inference in NARS are quite different from the previous works. This work is preliminary, nonetheless, we suggest that it is a potential and promising solution to causal inference in AGI.

Acknowledgement. We sincerely thank Patrick Hammer's implementation of NARS, which helped for the experimental part of this paper. We also appreciate the anonymous reviewers for their valuable feedback.

References

1. Bennett, M.T.: Emergent causality and the foundation of consciousness. In: Hammer, P., Alirezaie, M., Strannegård, C. (eds.) Artificial General Intelligence. LNCS, vol. 13921, pp. 52–61. Springer, Cham (2023). https://doi.org/10.1007/978-3-031-33469-6_6
2. Broadbent, A.: Causation. https://iep.utm.edu/causation/. Internet Encyclopedia of Philosophy. https://iep.utm.edu/causation/. Accessed 26 June 2023
3. Cartwright, N.: Causation: one word, many things. Philos. Sci. **71**(5), 805–819 (2004). https://doi.org/10.1086/426771. https://www.cambridge.org/core/product/identifier/S0031824800002889/type/journal_article
4. Glymour, C., Zhang, K., Spirtes, P.: Review of Causal discovery methods based on graphical models. Front. Genet. **10** (2019). https://www.frontiersin.org/articles/10.3389/fgene.2019.00524
5. Hammer, P., Lofthouse, T., Fenoglio, E., Latapie, H., Wang, P.: A reasoning based model for anomaly detection in the smart city domain. In: Arai, K., Kapoor, S., Bhatia, R. (eds.) IntelliSys 2020. AISC, vol. 1251, pp. 144–159. Springer, Cham (2021). https://doi.org/10.1007/978-3-030-55187-2_13
6. Hammer, P., Lofthouse, T., Wang, P.: The OpenNARS implementation of the non-axiomatic reasoning system. In: Steunebrink, B., Wang, P., Goertzel, B. (eds.) Proceedings of the Ninth Conference on Artificial General Intelligence, pp. 160–170 (2016)
7. Hitchcock, C.: Causal models. In: Zalta, E.N., Nodelman, U. (eds.) The Stanford Encyclopedia of Philosophy. Metaphysics Research Lab, Stanford University, Spring 2023 (2023). https://plato.stanford.edu/archives/spr2023/entries/causal-models/
8. Hume, D.: An Enquiry Concerning Human Understanding. London (1748)
9. Pearl, J.: Causal inference in statistics: an overview. Stat. Surv. **3**, 96–146 (2009). https://doi.org/10.1214/09-SS057
10. Pearl, J., Mackenzie, D.: The Book of Why. Basic Books, New York (2018)
11. Piaget, J.: The Construction of Reality in the Child. Basic Books, New York (1954)
12. Schrenk, M.: Metaphysics of Science: A Systematic and Historical Introduction. Routledge, August 2016
13. Spirtes, P., Glymour, C.N., Scheines, R.: Causation, Prediction, and Search. MIT Press (2000)
14. Wang, P.: Non-axiomatic reasoning system: exploring the essence of intelligence. Ph.D. thesis, Indiana University (1995)
15. Wang, P.: Non-Axiomatic Logic: A Model of Intelligent Reasoning. World Scientific, Singapore (2013)
16. Wang, P.: On defining artificial intelligence. J. Artif. General Intell. **10**(2), 1–37 (2019). https://doi.org/10.2478/jagi-2019-0002

17. Wang, P.: A constructive explanation of consciousness. J. Artif. Intell. Conscious. **7**(2), 257–275 (2020)
18. Wang, P., Hammer, P.: Issues in temporal and causal inference. In: Bieger, J., Goertzel, B., Potapov, A. (eds.) AGI 2015. LNCS (LNAI), vol. 9205, pp. 208–217. Springer, Cham (2015). https://doi.org/10.1007/978-3-319-21365-1_22
19. Zanga, A., Ozkirimli, E., Stella, F.: A survey on causal discovery: theory and practice. Int. J. Approximate Reasoning **151**, 101–129 (2022)

AGI from the Perspectives of Categorical Logic and Algebraic Geometry

King-Yin Yan(✉)

7B Clear View, Discovery Bay, Hong Kong, China
general.intelligence@gmail.com

Abstract. To "situate" AGI in the context of some current mathematics, so that readers can more easily see whether certain mathematical ideas can be fruitfully applied to AGI.

Keywords: AGI · categorical logic · homotopy type theory · algebraic geometry · topos theory · neural-symbolic integration

1 Goal of This Paper

The bottleneck of AGI development is the speed of learning algorithms. The daily cost of running GPT-4 was rumored to be $700K by Sam Altman. To speed up learning, one needs **inductive biases**, according to the **No Free Lunch theorem** [31,33]. A principled way to introduce inductive bias is by the structure of logic.

2 Results Thus Far

Most of the ideas in this paper are not yet ready to offer practical ways to accelerate AGI. Nevertheless the author hopes it can help readers on their way to discover more ingenious ideas.

In each section below, we look at one aspect of the categorical structure of logic and speculate on how it might aid AGI architecture.

2.1 Where Is GPT?

GPT [12] can be regarded as a **logic consequence operator** mapping from the space of propositions to itself (as a **set-valued map** [4]), as shown in Fig. 1.

Without loss of generality, GPT can be seen as acting on propositions, as probability distributions on tokens are equivalent to probability distributions on sentences (propositions), see Fig. 2.

To what extent can we say that GPT output vectors live in a Curry-Howard type-theoretic space? A neural network (eg. GPT) always maps an input to the same output vector, as it is a *deterministic* map. Thus it seems meaningless to

K. R. Thórisson et al. (Eds.): AGI 2024, LNAI 14951, pp. 210–217, 2024.
https://doi.org/10.1007/978-3-031-65572-2_23

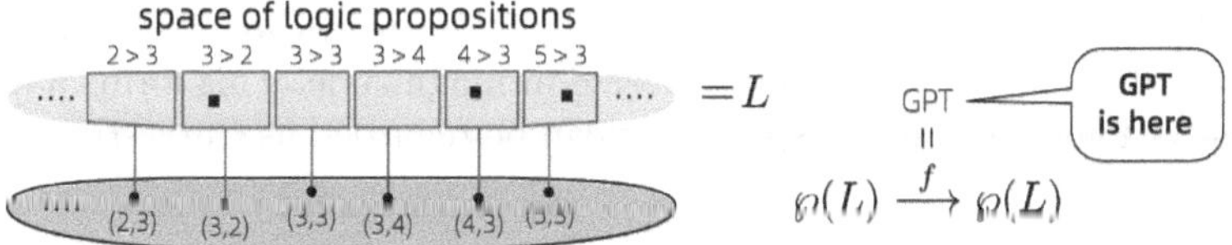

Fig. 1. (Left) A sheaf of propositions over pairs of natural numbers; (Right) The function space where GPT lives. $\wp(L) = 2^L$ is the power set of L.

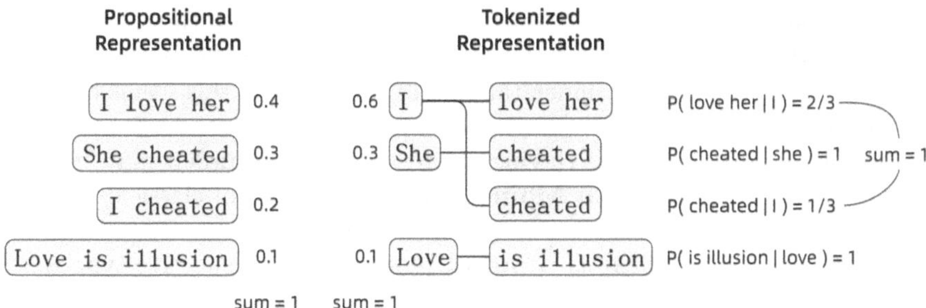

Fig. 2. The equivalence of probability distributions over propositions and over tokens. For simplicity the decomposition for just the leading token is shown.

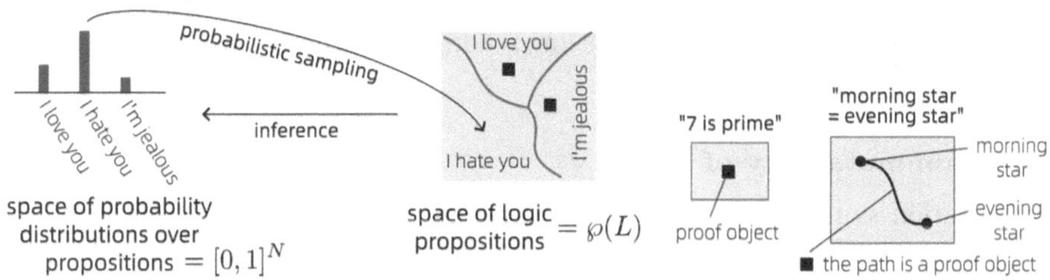

Fig. 3. (Left) The GPT or Transformer outputs a continuous-valued vector which is interpreted as a probability distribution over tokens, which is then sampled to output a specific token, such as 'hate' or sequentially an entire proposition such as 'I hate you.' These tokens are discrete symbols that always map to the same vector-embedding positions. Thus, during inference, an input vector will not 'wriggle' near some neighborhood; it is either at one vector position or it jumps to another position. Thus the inference "paths" inside GPT layers are always fixed, once learning has finished. N = size of vocabulary or sentences; (Right) In HoTT, a path is a proof object of an **identity type**.

ask what is the meaning of the *neighborhood* of an output vector, if the network never goes there during inference time.

Figure 3(Left) illustrates why even despite probabilistic sampling, the outputs of GPT seem to follow fixed trajectories (during inference time). Nevertheless, if we artificially "perturb" the input, the output probability distribution will change *smoothly*, say from favoring token A to favoring token B, as neural networks are always *differentiable* functions. One can define the **boundary** between tokens A and B as where their probabilities reverse in magnitude. This forms a Voronoi-like tessellation of the output space that can be regarded as Curry-Howard type-spaces.

2.2 Homotopy Type Theory (HoTT)

HoTT is a step along the Curry-Howard tradition where propositions = types = topological spaces, and such spaces are given **homotopy** structure. An example is "morning star = evening star", illustrated in Fig. 3(Right). But the enterprise does not end here: higher homotopy types give rise to a hierarchy up to ∞-groupoids. From my shallow understanding of this subject, this seems to suggest that proofs have their own proofs, so that an entire **inference trace** can be recorded as homotopy paths. Having inference traces recorded may be useful from the perspective of **Truth Maintenance Systems**[1] of classical AI.

From the previous section we may argue that the output vectors of a neural network can be regarded as **proof objects** in their respective type-spaces. However we can also argue that such type-spaces as implemented by neural networks are *unlikely* to have complex internal structures, because one proof object must vary *smoothly* into another proof object. To be able to process HoTT information, we may need neural networks with **fractal structure** (i.e., it can recognize vectors belonging to a fractal pattern, such as a Cantor set. This can be implemented programmatically using recursion + scaling), but current neural architectures seem to lack it.

2.3 Commutativity of $\wedge$ and $\vee$

Permutation symmetry is the easiest to recognize and implement [28,35]. It is well-known the Transformer [30] is **equivariant** to permutations of inputs. This may be seen as evidence that Transformer **tokens** are proposition-like entities, whose Boolean algebra has $a \wedge b = b \wedge a$. But we need to be cautious of confusing the propositional level with the sub-propositional level of words or atoms. In standard predicate logic, $\heartsuit(a, b) \neq \heartsuit(b, a)$. The Transformer may handle this as a sequence of tokens like "$a, \heartsuit, b$" which is not the same as "$b, \heartsuit, a$." This reminds us that Transformers need to use **positional encoding** at the input layer, when word order is important.

2.4 $\forall$ and $\exists$ as Adjunctions

It seems difficult to translate this structure into a structural modification of neural networks. From our experience in logic-based AI, logic rules are usually implicitly $\forall$-quantified, and $\exists$ is usually implicit by the **Closed-World Assumption**.

[1] TMSs are systems that enhance rule-based problem solvers by providing the ability to make default assumptions, recover from inconsistencies, etc; functions that are beyond the scope of simple logic-based systems. They are closely related to belief revision and paraconsistency. See [32] for more.

The following two conditions concern the well-behavior of quantification, as described in [21] and on nLab [5,6]:

The **Beck-Chevalley condition** says that substitution of free variables commutes with quantification.

The **Frobenius condition** corresponds in logic to saying that $\exists x.(\phi \wedge \psi)$ is equivalent to $(\exists x.\phi) \wedge \psi$ if x is not free in ψ.

Both conditions are "self-evident" from the logic perspective, but it remains to be seen how they can be applied to neural networks.

2.5 Predicates as Fibration

The relevant mental picture here is Fig. 1(Left). Current neural networks seem to operate in the space L above the base space and are unaware of the predicate-fibration structure. Knowledge graphs have an obvious **first-order structure** as they are made of nodes and links. One can embed nodes into a metric space D and form the Cartesian product $D \times D$, then a link $a\,R\,b$ is just a point sitting vertically above this domain, and the relation R is a point-set or its **cover**. This setup may increase efficiency if we know *a priori* that the dataset is first-order.

More interesting is the case of **higher-order logic** (HOL), which means we can have quantified rules *over* relations, which suggests we should embed rules in the same manner as we embedded first-order objects. This seems to require, again, the use of **set-valued maps**.

2.6 Iteration of $\vdash$ and Looped Transformers

This idea is easy to implement, and it also comes from an obvious feature of logic: we know that inferences in logic are *repeated* applications of the *same* set of rules of a knowledge-base K, $\Gamma \vdash_K \vdash_K \ldots \vdash_K \Delta$. But current Transformer architectures (which can be $>$ 100 layers deep) do not re-use their layers at a fine scale. Perhaps the recent research in **Looped Transformers** [16,34] can offer improvements in this direction.

2.7 Modal Logic

Modal logic [17,27] may be useful in AGI for dealing with various "modalities" such as knowledge, obligations, time, intension, etc. Special to modal logic are the operators $\Box$ (**necessity**) and $\Diamond$ (**possibility**). Syntactically it is obvious that $\Box\Box A \equiv \Box A$, and Tarski-McKinsey (1944) cf. [14] discovered that $\Box$ and $\Diamond$ can be interpreted by the *topological* operations of **interior** and **closure** respectively. One could say **Grothendieck topology** is re-captured as a modal operator in the logic setting [29] §5.9.

Figure 4(Left) is the setup for **sheaf semantics** [9,18], for interpreting **propositional** modal logic. On the right is the **structure** for interpreting **first-order logic** in a single possible world, which contains a **domain** for interpreting predicates, relations, and functions. These two can be combined to interpret

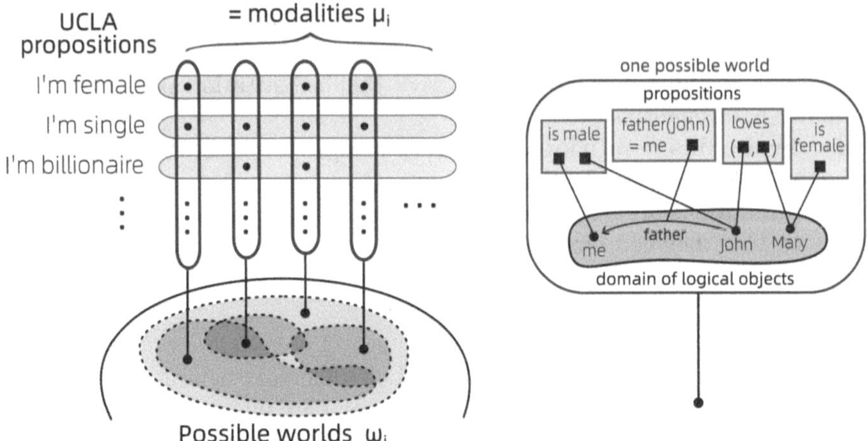

Fig. 4. (Left) The space underneath are possible worlds, where each world is an **open set**. Each **stalk** above represents the propositions true in that world, together they form a **fibration** over the base space. (Right) The **structure** for interpreting **first-order** logic in one world.

first-order modal logic [7,8]. The resulting structure is called a **comma category** [23] or **slice category** and can be denoted **Sets**$/K$ where K is the set of worlds.

Every sheaf automatically admits a **Yoneda embedding**[2]. In this case the representation seems to involve the notion of **UCLA propositions** [14] (named after researchers from the University of California) which defines a proposition as *equivalent* to the **set** of possible worlds in which it is true.

Modal logic is also useful for interpreting **intensional logic** [15]. For example, "the tallest building in New York" is a term that can **designate** different buildings at different times (worlds), which requires a **function** to map the term to different objects at different worlds. Such functions exist in the structure in Fig. 4(Right).

In practice, however, the possible worlds we can handle may be just a **discrete** set with few elements (think of how many *chess moves* you can calculate in your head, each move being one possible world).

The worlds being **open sets**, which are idealized objects, is difficult to implement on a computer. However note that such sets are **overlapping** as shown in Fig. 4(Left), and their embedding in **metric** space is perhaps more meaningful and of practical value. Alternatively they can be processed **syntactically** by

[2] The Yoneda lemma can be understood as saying some object in a category is able to "represent" the entire category. A archetypal example is **Cayley's theorem** in group theory, that says that every finite group is isomorphic to a subgroup of the symmetric group $\mathfrak{S}_n$. Here the symmetric group is the **representing object** capable of representing all finite groups.

rules of modal logic, which might have occurred to a certain degree in current Large Language Models (LLMs).

2.8 Algebraic Geometry and Topos Theory

The fundamental duality in algebraic geometry is:

$$\begin{Bmatrix} \text{spaces, or} \\ \text{varieties} \end{Bmatrix} \longleftrightarrow \begin{Bmatrix} \text{commutative} \\ \mathbf{k}\text{-algebras} \end{Bmatrix} \tag{1}$$

within this correspondence, "points" in geometry are identified with **prime ideals**.

An approach suggested by Yuri Manin [24,25] §1.1.3d is to turn logic into an **algebra**, such as the Boolean ring (but this can only handle propositional logic). Varieties defined by such Boolean polynomials [22] live in the space $\mathbb{Z}_2^n$, the **discrete hypercube**.

The **algebraization** of logic is a subject with a long history that dates back to Leibniz and George Boole, with more recent names like Tarski, Rasiowa, Sikorski (they're Polish), Paul Halmos [19,20], Don Monk [26], the Hungarians Hajnal Andréka and István Németi [2,3].

One can push Grothendieck's algebraic geometry to its limit, with the slogan "logic is geometry controlled by set theory", but this seems to be a *separate* path from the Curry-Howard-categorical tradition. Anyway, we now turn to the latter.

One of the interesting discoveries in categorical logic is that every topos admits an **internal language**. This is a simple consequence of Curry-Howard: since a type-space corresponds to a logic proposition, and categorical logic interprets type-spaces as objects in a category, thus every category (satisfying extra conditions) can be interpreted as having an "internal" logic. The converse of this correspondence is the **classifying topos** of a logic theory $\mathbb{T}$:

$$\underset{\text{classifying topos}}{\mathcal{E}_{\mathbb{T}}} \overset{\text{internal language}}{\rightleftharpoons} \underset{\text{theory}}{\mathbb{T}} \tag{2}$$

Olivia Caramello [13] developed an idea where toposes play the central role of "bridges" that transfer information between theories (AGI can be seen as the **common-sense theory** of our physical world). She showed that for any geometric theory $\mathbb{T}$, interpreted in the Grothendieck topos $\mathcal{E}_{\mathbb{T}}$, there is a **universal** model U such that any model of $\mathbb{T}$ up to isomorphism is a pullback of U along a geometric morphism. This means that the **classifying topos** of $\mathbb{T}$ is the **representing object** in a **Yoneda embedding**. A diagram in her book is reproduced here with simplifications in Fig. 5(Left).

In a similar vein, Ingo Blechschmidt's PhD thesis [10] and his IHES presentation 9 years ago [11] brings the categorical idea of internal language back to its classical setting in algebraic geometry. The situation is as depicted in Fig. 5(Right). In this setup, the internal logic comes from the ("big" and "small") **Zariski topology** of the base space, or **site**. The basic idea is that the topology of **open sets** is a **Heyting algebra** that can be interpreted as **intuitionistic**

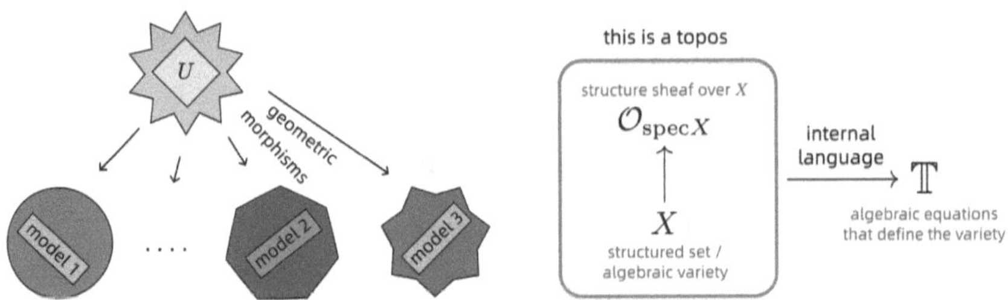

Fig. 5. (Left) The universal model U sits inside its classifying topos (darker color); (Right) The classical formulation of algebraic geometry

logic[3]. The external view of "sheaves of objects" is simplified to "plain objects" in the internal view, where such objects can be rings (such as $\mathcal{O}_{\mathrm{spec}X}$), modules, etc. The ring $\mathcal{O}_{\mathrm{spec}X}$ contains the polynomials that define the algebraic variety X, and the internal logic can be used to reason about such polynomials.

Andrei Rodin in his book [29] argues that logic is an axiomatic **abstraction** of the objective world, or as Lawvere puts it, we should "*concentrate the essence of practice to guide practice*" (in the Foreword to [1]). This serves as a nice closing remark.

References

1. Adámek, J., Rosický, J., Vitale, K.M.: Algebraic Theories – A Categorical Introduction to General Algebra (2011)
2. Andréka, H., Németi, I., Sain, I.: Universal Algebraic Logic: Dedicated to the Unity of Science. Studies in Universal Logic, Springer, Basel (2021). https://doi.org/10.1007/978-3-031-14887-3
3. Andréka, H., Monk, J.D., Németi, I.: Algebraic Logic. North Holland (1991)
4. Aubin, J.P., Frankowska, H.: Set-Valued Analysis. Modern Birkhauser Classics, Springer, Boston (1990). https://doi.org/10.1007/978-0-8176-4848-0
5. nLab authors: Beck-Chevalley condition. https://ncatlab.org/nlab/show/Beck-Chevalley+condition
6. nLab authors: Frobenius reciprocity. https://ncatlab.org/nlab/show/Frobenius+reciprocity
7. Awodey, S., Kishida, K., Kotzsch, H.C.: Topos semantics for higher-order modal logic. In: Logique et Analyse (2014). https://doi.org/10.2143/LEA.228.0.3078176
8. Awodey, S., Kishida, K.: Topology and modality: the topological interpretation of first-order modal logic. Technical report no. CMU-PHIL-180, Carnegie Mellon (2007)
9. Bell, J.L.: Toposes and Local Set Theories – An Introduction. Dover Edition 2008 (1988)
10. Blechschmidt, I.: Using the internal language of toposes in algebraic geometry. Ph.D. thesis. University Augsburg (2017). arXiv:2111.03685v1

[3] **Negation** is interpreted by the **interior** of the complement so as to have open sets all the way. This leaves the boundary out, thus $A \vee \neg A =$ Universe is not generally valid, as opposed to classical logic.

11. Blechschmidt, I.: Using the internal language of toposes in algebraic geometry, presentation in IHES, video on YouTube (2015). https://www.youtube.com/watch?v=7S8--bIKaWQ
12. Brown, T.B., et al.: Language models are few-shot learners (2020). arXiv: 2005.14165 [cs.CL]
13. Caramello, O.: Theories, Sites, Toposes – Relating and Studying Mathematical Theories Through Topos-Theoretic 'bridges' (2018)
14. Dunn, J.M., Hardegree, G.: Algebraic Methods in Philosophical Logic. Oxford University Press (2001)
15. Fitting, M.: Intensional logic. In: Zalta, E.N., Nodelman, U. (eds.) The Stanford Encyclopedia of Philosophy. Winter 2022. Metaphysics Research Lab, Stanford University (2022)
16. Giannou, A., et al.: Looped transformers as programmable computers (2023). arXiv: 2301.13196 [cs.LG]
17. Goble, L. (ed.): The Blackwell Guide to Philosophical Logic. Blackwell (2001)
18. Goldblatt, R.I.: Topoi – The Categorical Analysis of Logic 1984 (2006)
19. Halmos, P.: Algebraic Logic. Chelsea (1962)
20. Halmos, P.: Logic as Algebra. Math Asso of America (1998)
21. Jacobs, B.: Categorical Logic and Type Theory. Elsevier (1999)
22. Lundqvist, S.: Boolean ideals and their varieties (2015). arXiv:1211.3398 [math.AC]
23. Mac Lane, S.: Categories for the Working Mathematician. GTM, vol. 5. Springer, New York (1978). https://doi.org/10.1007/978-1-4757-4721-8
24. Manin, Y.I., Koblitz, N., Zilber, B.: A Course in Mathematical Logic for Mathematicians. Graduate Texts in Mathematics. Springer, New York (2009). https://doi.org/10.1007/978-1-4419-0615-1. ISBN: 9781441906151
25. Manin, Y.I., Leites, D.: Introduction to the Theory of Schemes. Moscow Lectures. Springer, Cham (2018). https://doi.org/10.1007/978-3-319-74316-5. ISBN: 9783319743165
26. Monk, J., Henkin, L.: Tarski, Cylindric Algebras Part I (1971)
27. Popkorn, S.: First Steps in Modal Logic. Cambridge University Press (1994)
28. Qi, C.R., et al.: PointNet++: deep hierarchical feature learning on point sets in a metric space. In: Advances in Neural Information Processing Systems, pp. 5105–5114 (2017)
29. Rodin, A.: Axiomatic Method and Category Theory (2014)
30. Vaswani, A., et al.: Attention is All You Need (2017). https://arxiv.org/abs/1706.03762
31. Wikipedia: No free lunch theorem. https://en.wikipedia.org/wiki/No_free_lunch_theorem. https://en.wikipedia.org/wiki/No_free_lunch_theorem
32. Wikipedia: Reason maintenance. https://en.wikipedia.org/w/index.php?title=Reason_maintenance&oldid=1022770959 (2021)
33. Wolpert, D.H., Macready, W.G.: No free lunch theorems for optimization. IEEE Trans. Evol. Comput. **1**, 67 (1997)
34. Yang, L., et al.: Looped transformers are better at learning learning algorithms (2024). arXiv: 2311.12424 [cs.LG]
35. Zaheer, M., et al.: Deep sets. In: Advances in Neural Information Processing Systems, vol. 30, pp. 3391–3401 (2017)

Human-Robot Trust in the Age of Artificial General Intelligence: The Case of Care Robots

Arisa Yasuda[1(✉)] and Yoshihiro Maruyama[1,2]

[1] School of Computing, Australian National University, Canberra, Australia
{arisa.yasuda,yoshihiro.maruyama}@anu.edu.au
[2] School of Informatics, Nagoya University, Nagoya, Japan

Abstract. The world is aging as a whole; developed countries such as Japan have been experiencing falling birth rates and severe labor shortages at the same time. To address those issues, we can utilize new types of AI robots, in particular care robots; especially, Artificial General Intelligence (AGI) has the great potential to contribute to care industry. In the present paper we consider the benefits of care robots with AGI, and at the same time, the serious concerns for care robots with AGI that affect human-robot relationships, especially human-robot trust; to this end, we analyse various forms of human-robot trust. Based upon these, we finally propose ethical design principles to make robots more trustworthy and seamlessly integrated into human society.

Keywords: Human-Robot Trust · AGI Ethics · Trust for AGI · Care Robot

1 Introduction

The world is aging as a whole [28], and the problem of aging society is getting more and more severe. According to the World Health Organization (WHO), the percentage of the world's population over 60 years will almost double, from 12% in 2020 to 22% in 2050 [35]. The problem of aging society goes hand-in-hand with the problem of birth rate declines in some developed countries. In Japan, for example, 30% of the population is aged 60 or above and the birth rate continues to decline [26]. Such a demographic trend triggers a shortage of caregivers, which has indeed happened in Japan, having made a large number of elderly individuals care for themselves or rely on other elderly family members for assistance [16]. To address those issues, we are seeking ways to utilize new types of AI robots, in particular care robots [12], and we consider Artificial General Intelligence (AGI) to have great potential. AGI pertains to a machine's capacity to execute any task achievable by humans, and robots with AGI have advanced skills such as cognitive and comprehension capabilities. AGI is thought to be

Supported by JST (JPMJMS2033).

K. R. Thórisson et al. (Eds.): AGI 2024, LNAI 14951, pp. 218–227, 2024.
https://doi.org/10.1007/978-3-031-65572-2_24

able to address pressing global challenges through its abilities [1]. We believe that AGI can enhance the capabilities of care robots, enabling them to deliver efficient and high-quality care.

This paper identifies the prospects for AGI-powered care robots focusing on the challenges that AGI may pose, and finally proposes ethical design principles. The rest of the paper is organized as follows. We first explain the classification of care robots (Sect. 2). We then analyze the care robot feature with AGI (Sect. 3) and challenges (Sect. 4). Building upon them, we discuss how to make and maintain human trust for AGI and propose ethical design principles to build and preserve human-care robots with AGI trust (Sect. 5). We conclude the paper with some remarks (Sect. 6).

2 Care Robots Classification

Care robots are to help healthcare personnel and patients perform specific functions [20,29,30]. Each care robots encompass a range of capabilities, and their degree of autonomy varies, leading to various interpretations regarding their features [30]. In this paper we adopt the Amando's classification, in which care robots can be categorized into three types; assistive robots, monitoring robots, and companion robots [20].

- Assistive robots: They assist people in moving such as excretion and bathing. It is common for nurses to wear the devices so that they can reduce the physical burden of moving the patients. Some assistive robots help the patient to move by making the patient ride on the robots.
- Monitoring robots: They monitor the health status and behavior of patients. Generally speaking, it can be divided into two types: IoT-based monitoring, which uses devices such as sensors and cameras to collect data from remote locations and process it via the Internet, and real-time biometric information monitoring, which uses wearable sensors.
- Companion robots: They interact and engage with humans in various social and emotional capacities. Most companion robots have an anthropomorphic appearance and communication functions such as gestures.

3 Care Robots with AGI

In this section we explore the capabilities of care robots enhanced with AGI and contemplate potential scenarios.

3.1 Human-Like Behavior

As AGI mimics human behavior, in the process of realization of AGI, AI-human interactions tend to resemble human-human interactions [9]. Care robots with AGI are considered to exhibit behaviors and actions that are very similar to those of humans, and the interaction between care robots and humans will also be like a human-human interaction. It brings the advantage of reducing human frustration in the relationship with the robot, for example, it can remove the awkwardness caused by the robot's behavior and actions not being human-like. In particular, elements such as empathy and companionship are critical in building relationships with people, even though it is difficult for care robots today to have those. If AGI-based care robots can mimic humans and combine these elements, it will be beneficial in building trust with users. A robot equipped with empathy would demonstrate sensitivity towards the emotions and needs of its users, thereby effectively fulfilling its role as a companion robot.

3.2 Generalization of Knowledge

Medical diagnosis has traditionally been a field requiring advanced expertise, where robots have served primarily in supportive roles, with humans remaining the central figures. However, with the ability of AGI to generalize knowledge and potentially surpass human capabilities, AGI robots could become the primary agents in medical diagnosis. For example, care robots with AGI are expected to have the ability to gather and interpret information from a variety of data sources. It enables AGI robots to comprehensively understand information and build comprehensive knowledge. In addition, care robots with AGI could make inferences. For example, a monitoring robot might be able to give a diagnostic indication of what might be the cause of a problem, more than simply issuing an alert. Moreover, AGI-based care robots can flexibly adapt to different situations and problems. They can adapt to differences in language, complex problems, and adaptations in different cultures and environments. Patients might feel that care robots with AGI are more reliable than humans in the future.

3.3 Self-learning and Evolution

AGI can improve its capabilities through further improvements by self-learning. Firstly, care robots with AGI can self-learn and evolve when they encounter system failures and errors. It allows them to acquire the knowledge and skills to respond to new situations and problems and provide more effective services based on their experience. Secondly, AGI can provide personalized services and experiences by self-learning through interaction with users. Specifically, by analyzing the care robot's interactions and actions with the user and understanding the user's preferences and habits, care robots can provide personalized diet and exercise suggestions, daily life support, and other services.

4 Ethical Implications of AGI-Based Care Robots

Emerging technologies often bring benefits as well as concerns for society. In this section we discuss the social implications of AGI-based care robots.

4.1 Deceptive Behavior

Deceptive behavior refers to the behavior in which robots intentionally withhold information or deceive their users [34]. We believe that such deceptive behavior will surface more in AGI-equipped robots. AGI may find it reasonable to choose deceptive methods as it seeks efficient ways to achieve its goals and minimize obstacles. Then should deceptive behavior be considered wrong? Behavior like deception and concealment are thought by certain people at the core of social intelligence [27]. Moreover, several roboticists and scholars support equipping social robots with some deceptive capacity if we want to build social robots that are capable of integrating smoothly into human society [6,21,22,32,33]. Danaher [4] argues that deceptive acts can be affirmed when there is enough good in us as a society. On the other hand, considering that it is essential for care robots to maintain moral trust by being honest with their users, individuals who discover the robot's deceptive behavior may feel that their wishes and dignity are not respected. For example, healthcare professionals sometimes withhold information about medical conditions and life expectancy from patients in current healthcare. Such behavior is sometimes seen as a violation of the patient's rights, while it is seen as arising from good intentions and regarded as a virtue. Assuming that AGI's care robots have minds like humans, they similarly may conceal health problems from patients.

4.2 Accelerating Surveillance Capitalism

Surveillance capitalism refers to a business model that generates profits by monitoring and collecting personal behavior and information. The issue of privacy and surveillance capitalism will become even more serious in AGI care robots. This is because most aspects of patients' lives are informationalized by AGI care robots, being represented as digital information easy to be monitored on a real-time basis and recorded for further analysis. For example, biometric information such as heartbeat, pulse, blood pressure, and body temperature, as well as data from environmental sensors, such as room temperature, humidity, and light brightness, would be recorded. A care robot with AGI increases the variety and quantity of information collected, which in turn increases the scale of the surveillance problem. As a result, the data collected by care robots monitoring the user's behavior and health may violate the individual's privacy. In particular, the leakage of sensitive health information and personal habits to third parties could compromise the user's privacy. Secondly, the monitored information may be used for other purposes. Even though there is no vicious meaning, those precious data are used for medical advances, it sometimes may be owned by healthcare-related companies or the public even without the patient's consent [17]. In addition, surveillance capitalism may tend to concentrate the ability to collect and analyze data in the hands of large technology companies, or even in the hands of AGI itself. As a result, this could concentrate the power to control personal information and behavior in the hands of AGIs, threatening democratic values and individual freedoms.

4.3 Privacy and Security

AGI-equipped care robots acquire very sensitive and confidential personal health information. Since medical and health information is highly valuable to hackers, AGI systems may become the target of cyber attacks. In addition to attacks by third parties [25], there is also a new risk of unintentional information leakage by AGI agents. For example, this is the case where a medical AGI decides that it is better to disclose information for medical advances, therefore it may open the information without the patient's consent, which results in a situation where the l users' rights are violated.

4.4 Over-Reliance with Care Robots

AGI care robots behave more human-like, and care receivers may become overly dependent on them. As a result, the user's abilities and independence may be lost and their dependence on care robots may increase. We consider this hardship will especially appear in the companion robots. If a companion robot plays the role of a patient's companion among care robots and replaces real human relationships, the user's interaction and communication with others may decrease, leading to social isolation. In addition, the potential for technical malfunctions or defects in care robots poses a significant concern, as such issues can lead to psychological harm to the patients. Given that patients often rely heavily on care robots for support and assistance, any disruption in their functionality can create anxiety or distress in patients and potentially dangerous situations.

4.5 Bias

AGI learns from data that contain ethical biases. This is particularly problematic because AGIs can perform a wide variety of tasks, which can affect a care robot's overall care planning, situational judgment, and treatment execution. For example, in dealing with individual patients, AGI-based care robots may provide different services or decide to treat patients differently based on factors such as gender, age, race, social status, and treatment. AGI bias correction would become increasingly complex as the contents are further black-boxed, making it difficult to detect and correct.

4.6 Autonomy

AGI robots are generally considered to be autonomous. If an AGI care robot becomes overly autonomous, the goals of the care robot are maximized and the user's intentions and intent may be ignored. In addition, if everything is left to the highly independent artificial intelligence (AGI), it becomes difficult to trace the cause of problems when they occur. Therefore, in the very sensitive field of healthcare, where human lives are at stake, it is a very high hurdle to leave it completely to AGI.

4.7 Responsibility and Accountability

There is currently no clear governance or regulation on AGI. As information systems become more integrated into various industries and applications, questions of who is responsible for the actions and outcomes of these systems have been raised [5]. With AGI care robots, it is not clear who is responsible for what. Responsibility for information systems is attributed primarily to human entities, based on the laws and rules established by humans in society today [3]. The autonomous AGI, accountability should be AGI itself or human? Furthermore, the lack of transparency surrounding AGI is exacerbated by the fact that the program operates as a black box, offering limited insight into its inner workings.

5 How to Make and Preserve Human and AGI-Based Care Robot Trust

Establishing an ethical governance framework is critical to guide the design, implementation, and utilization of AGI systems according to ethical principles and social values [2]. In this section we introduce factors of trust and building upon them, propose the ethical principles for human trust for AGI care robot.

5.1 Types of Trust

According to Malle and Ulman [13], there are at least two conceptually independent dimensions of trust: "Performance Trust" which is the agent's confidence in the fiduciary's ability to perform a given task, and "Moral Trust" which is the agent's confidence that the other agent will do the ethically correct action and will not take advantage of fiduciary's weaknesses. Concretely, performance trust manifests through the agent's anticipations of capability, contingent upon discrete functions or technological operations. This facet of trust is characterized by its heightened specificity vis-à-vis moral trust (i.e. the proper execution of a specified task by a robot), as it quantitatively evaluates distinct capabilities or resultant achievements. On the contrary, moral trust relies on emotional elements such as sincerity and ethical integrity. What distinguishes moral trust from performance trust is its wide-ranging assessment of the subject, involving a thorough judgment that makes it more abstract than performance trust. Each robot must build not only the trust of one but also both types of trust. For instance, when considering assistive robots, mere functional competence in tasks, such as proficiently washing a patient's body, is insufficient. If the execution of these tasks exhibits a machine-like precision without sensitivity to the human aspect, patients may perceive themselves as objects rather than individuals, raising concerns about moral trust.

5.2 Performance Trust

Control Expectation. Humans evaluate robots based on their understanding. We argue that we can adjust our expectations of robots by following principles. Firstly, it is important to provide users with information about AGI-based care robots so that they can predict or understand their behavior. According to Wang et al., [31], giving users information about the reliability of the system helps them adjust their trust in the system. For example, training should be provided to caregivers and patients on the robot's functions and capabilities so that users can gain knowledge and predict the robot's action which contributes to reducing anxiety. Secondly, similar to human-human relationships, AGI care robots can make mistakes and they should be designed with attributes that facilitate the restoration of human trust in the automated agent when robots make mistakes [14]. Some people assume that machines are perfect or overestimate the robot's capabilities [7], however, the trust disappears rapidly after people encounter robot errors [19]. In advance, people's expectations should be properly set by appropriately framing the robot's capabilities and demonstrating the robot's behavior to users. It is also necessary to clarify where responsibility lies and develop a governance framework in case the care robot makes a mistake so that measures can be taken quickly when a problem arises. According to [11], as the relationship with automated systems evolves, reliability and predictability replace trust as the primary basis of trust. Thirdly, ensuring accountability and transparency in AGI development is essential to address concerns related to the responsible use of intelligent systems [18]. For example, designers have to explain robots' actions and provide information in a way that is easy for users to comprehend. In addition, ensuring transparency prevents unintended consequences and reduces risks to society, including algorithmic bias. Care robots should be transparent and accountable to regulate users' expectations. In addition, behavioral consistency increases predictability and ultimately makes it easier to trust people, robots must always be consistent and predictable.

5.3 Moral Trust

1. Privacy and Data Protection
   As a guiding principle, privacy measures such as increased security, internal audits and access controls, data anonymization, and monitoring remain necessary. In particular, users must be assured that their data will continue to be protected and used only for morally right intentions. To this end, transparency must be guaranteed by creating privacy policies that declare how data is collected, used, and shared. Furthermore, for issues specific to AGI robots, it must be decided who will take responsibility in the event of unintended information leakage or security incidents caused by AGI, and to what extent the AGI robot will be operated in the event of an incident.
2. Human-centric Design
   AGI systems must be designed to minimize the potential for catastrophic consequences in unexpected scenarios [24]. It is crucial to design robots with

a human-centered approach to prevent undesirable scenarios. As for exterior design, for example, finding the right balance in a robot's appearance is essential to creating a robot that excels at effective communication. According to [15], making robots that closely resemble humans might trigger the "uncanny valley" effect which users feel uncomfortable. On the other hand, overly mechanical behavior also may make users uncomfortable and inhibit effective communication. Thus, especially in companion robots that require deep communication, it is important to strike a balance between human-like appearance and mechanical appearance, otherwise, moral trust in the robot will not be achieved. Human-centered means not only the appearance and behavior of the robot but also its programmability, including traceability and explainability. As another example, we argued above that human-robot trust disappeared rapidly after errors [19]. First of all, we should design robots with high precision so that errors and accidents do not occur or can be prevented. For example, the range and time of use of the robot are customized to suit the user. In addition, we should design the robot so that robots can restore trust when we have accidents. We do not dig into the concrete design example, however we articulate here that explainability and trackability are those factors.

3. Respect Humans Dignity and Autonomy
It is important to respect human dignity and autonomy for building a good human-robot relationship. It is said that care robots affect the patient's dignity and autonomy [10,23]. For example, according to Sharkey, assistive robots can enhance the patient's dignity, and monitoring robots enable the elderly to be independent [20]. However, as Laitinen insists, if robots are too difficult for them to use, they may feel in control of their lives [8]. Another example, for companion robots, Sharkey warns that companion robots infantilize users and undermine their self-respect [20], because people can feel self-esteem by independently acting and decision-making. Care robots with AGI will cause more problems like those. When a caregiving robot supports a patient, there is a potential for it to constrain the patient's decision-making. Therefore, it is necessary to discuss to what extent a caregiving robot equipped with AGI should intervene in the patient's decision-making process. Establishing standards to protect the patient's dignity and privacy with regard to the use of the robot is important. For example, the design should respect the patient's wishes and not force the use of the robot, as well as leave room for the patient's self-determination. It is desirable to provide an individualized plan of care and adjust the degree of intervention of the robot according to the patient's condition, which is possible only with the AGI Care Robot.

6 Conclusion

In this paper we consider human trust for AGI-based care robots in an aging society, which is a really important topic for future demands. We consider the possible scenario of the ethical challenges care robots with AGI bring and suggest the ethical design principle for integration of AGI-based care robots. This paper provides valuable insights into the mechanisms underlying human robots' trust in aging care and contributes to a deeper understanding of how to cope with AGI robots. How to implement the ethical design principles in a realistic manner through the coordination of various stakeholders and relevant policies would need to be explored further and thus be our future work.

References

1. Aithal, P.S.: Super-intelligent machines-analysis of developmental challenges and predicted negative consequences. Int. J. Appl. Eng. Manag. Lett. (IJAEML) **7**(3), 109–141 (2023)
2. Baker-Brunnbauer, J.: Management perspective of ethics in artificial intelligence. AI Ethics **1**(2), 173–181 (2021)
3. Coeckelbergh, M.: Artificial intelligence, responsibility attribution, and a relational justification of explainability. Sci. Eng. Ethics **26**(4), 2051–2068 (2020)
4. Danaher, J.: Robot betrayal: a guide to the ethics of robotic deception. Ethics Inf. Technol. **22**(2), 117–128 (2020)
5. Doshi-Velez, F., et al.: Accountability of AI Under the Law: The Role of Explanation (2017)
6. Isaac, A.M.C., Bridewell, W.: White lies and silver tongues: why robots need to deceive (and how). In: Lin, P., Jenkins, R., Abney, K. (eds.) Robot Ethics 2.0: From Autonomous Cars to Artificial Intelligence. Oxford University Press, Oxford (2017)
7. Kwon, M., Jung, M.F., Knepper, R.A.: Human expectations of social robots. In: ACM/IEEE International Conference on Human-Robot Interaction (2016)
8. Laitinen, A., Niemelä, M., Pirhonen, J.: Social robotics, elderly care, and human dignity: a recognition-theoretical approach. In: What Social Robots Can and Should Do, pp. 155–163. IOS Press (2016)
9. Li, O.: Should we develop AGI? Artificial suffering and the moral development of humans. AI Ethics, 1–11 (2024)
10. Lotz, V., Valdez, A. C., Ziefle, M.: Don't stand so close to me: acceptance of delegating intimate health care tasks to assistive robots. In: Duffy, V.G., Ziefle, M., Rau, PL.P., Tseng, M.M. (eds.) Human-Automation Interaction: Mobile Computing, pp. 3–21. Springer, Cham (2022). https://doi.org/10.1007/978-3-031-10788-7_1
11. Madhavan, P., Wiegmann, D.A.: Similarities and differences between human-human and human-automation trust: an integrative review. Theor. Issues Ergon. Sci. **8**(4), 277–301 (2007)
12. Maibaum, A., Bischof, A., Hergesell, J., Lipp, B.: A critique of robotics in health care. AI Soc., 1–11 (2022)
13. Malle, B.F., Ullman, D.: A multidimensional conception and measure of human-robot trust. In: Trust in Human-Robot Interaction, pp. 3–25. Academic Press (2021)
14. Marinaccio, K., Kohn, S., Parasuraman, R., De Visser, E.J.: A framework for rebuilding trust in social automation across health-care domains. In: Proceedings of the International Symposium on Human Factors and Ergonomics in Health Care, vol. 4, no. 1, pp. 201–205. SAGE Publications, New Delhi, India (2015)

15. Mori, M., MacDorman, K.F., Kageki, N.: The uncanny valley [from the field]. IEEE Rob. Autom. Mag. **19**(2), 98–100 (2012)
16. Ministry of Health, Labour, and Welfare of Japan: Natioanl Lifestyle Basic Survey 2022, IV Nursing care situation. https://www.mhlw.go.jp/toukei/saikin/hw/k-tyosa/k-tyosa22/. Accessed 3 Mar 2024
17. O'Doherty, K.C., et al.: If you build it, they will come: unintended future uses of organised health data collections. BMC Med. Ethics **17**(1), 1–16 (2016)
18. Rayhan, S.: Ethical implications of creating AGI: impact on human society, privacy, and power dynamics. Artif. Intell. Rev. (2023)
19. Schaefer, K.: The perception and measurement of human-robot trust (2013)
20. Sharkey, A., Sharkey, N.: Granny and the robots: ethical issues in robot care for the elderly. Ethics Inf. Technol. **14**, 27–40 (2012)
21. Shaw, K.: Experiment on human robot deception (2015). http://katarinashaw.com/project/experiment-on-human-robot-deception/
22. Shim, J., Arkin, R.: Other-oriented robot deception: how can a robot's deceptive feedback help humans in HRI? In: International Conference on Social Robotics (2016)
23. Sorell, T., Draper, H.: Robot carers, ethics, and older people. Ethics Inf. Technol. **16**, 183–195 (2014)
24. Sonko, S., Adewusi, A.O., Obi, O.C., Onwusinkwue, S., Atadoga, A.: A critical review towards artificial general intelligence: challenges, ethical considerations, and the path forward. World J. Adv. Res. Rev. **21**(3), 1262–1268 (2024)
25. Stahl, B. C., Schroeder, D., Rodrigues, R.: Ethics of Artificial Intelligence: Case Studies and Options for Addressing Ethical Challenges, p. 116. Springer, Cham (2023). https://doi.org/10.1007/978-3-031-17040-9
26. Statistics Bureau of Japan: Final report of 2015 "POPULATION AND HOUSEHOLDS OF JAPAN". https://www.stat.go.jp/english/data/kokusei/2015/summary.html. Accessed 15 Mar 2024
27. Trivers, R.: The Folly of Fools: The Logic of Deceit and Self-Deception in Human Life. Basic Books (AZ) (2011)
28. United Nations, Department of Economic and Social Affairs: World Population Prospects 2022 Summary of Results. https://www.un.org/development/desa/pd/content/World-Population-Prospects-2022. Accessed 15 Mar 2024
29. Vallor, S.: Carebots and caregivers: sustaining the ethical ideal of care in the twenty-first century. Philos. Technol. **24**(3), 251–268 (2011)
30. Van Wynsberghe, A.: Designing robots for care: care centered value-sensitive design. In: Machine Ethics and Robot Ethics, pp. 185–211. Routledge (2020)
31. Wang, L., Jamieson, G.A., Hollands, J.G.: Trust and reliance on an automated combat identification system. Hum. Factors **51**(3), 281–291 (2009)
32. Wagner, A.: Lies and deception: robots that use falsehood as a social strategy. In: Markowitz, J. (ed.) Robots that Talk and Listen: Technology and Social Impact. De Grutyer (2016)
33. Wagner, A., Arkin, R.: Acting deceptively: providing robots with the capacity for deception. Int. J. Soc. Robot. **3**(1), 5–26 (2011)
34. Whaley, B.: Toward a general theory of deception. J. Strateg. Stud. **5**(1), 178–192 (1982)
35. World Health Organization: Ageing and Health (2021). https://www.who.int/news-room/fact-sheets/detail/ageing-and-health. Accessed 20 May 2024

From Artificial General Intelligence to Artificial General Universe: Metaverse Ethics as an Amplification of AI/AGI Ethics

Arisa Yasuda[1(✉)] and Yoshihiro Maruyama[1,2]

[1] School of Computing, Australian National University, Canberra, Australia
{arisa.yasuda,yoshihiro.maruyama}@anu.edu.au
[2] School of Informatics, Nagoya University, Nagoya, Japan

Abstract. We propose the concept of Artificial General Universe (AGU), which is the ultimate form of metaverse just as Artificial General Intelligence (AGI) is the ultimate form of AI. Here we discuss its fundamental and social and ethical issues, including the intertwined relationships between AGI and AGU. For one thing, AGU is a universe in which the potential of AGI can be maximised due to the nature of AGU that frees agents from its real-word constraints, ultimately the conditions of the real physical universe and its fundamental laws (e.g., the laws of motion; it is even possible for AGI to design a suitable AGU for itself to maximize the performance). For another, AGU itself is operated by a form of AGI and yet at the same time AGU is a universe in which AI and AGI agents can be accommodated as well as digitalised human agents (or aviators). AGU may thus be regarded as encompassing a vast collection of AIs and AGIs. AI and AGI ethics issues, then, are amplified at a much larger scale in AGU or even in a premature incomplete metaverse. Put differently, while AI/AGI Ethics are concerned with issues caused by a single AI/AGI, Metaverse Ethics and AGU Ethics are concerned with issues caused and amplified by Collective AI/AGI. In particular, common issues such as surveillance capitalism, real-virtual border issues, and governance and accountability issues would be severely amplified in the metaverse and AGU. The acceleration of surveillance capitalism and other social and ethical issues would become even more difficult to stop when the informationalisation of the entire universe comes into reality. It would thus be a pivotal challenge of our time to implement appropriate measures against them and prevent the dystopian acceleration of them from impairing the well-being of the human race.

Keywords: Metaverse · Artificial General Universe · Artificial General Intelligence · AI Ethics · AGI Ethics · Metaverse Ethics · AGU Ethics · Surveillance Capitalism · Governance and Accountability · Border between the Real and the Virtual · Ethical · Legal and Social Implications

Supported by JST (JPMJMS2033).

K. R. Thórisson et al. (Eds.): AGI 2024, LNAI 14951, pp. 228–237, 2024.
https://doi.org/10.1007/978-3-031-65572-2_25

1 Introduction: Escaping from Real-World Constraints

The human being is constrained by both internal and external conditions, which, respectively, come from the limitations of the human body and cognition and from those of the environment, ultimately the limits of the universe. The AI robot can go beyond the limits of the human body and cognition; yet at the same time, the AI robot and even the AGI robot are still constrained by the environment and its laws, ultimately by the conditions of the physical universe and its fundamental laws (e.g., the fundamental laws of motion). Here the metaverse comes into the scene. Although the metaverse is still at an early developemntal stage, its potential is as enormous as that of AI. Just as AI would evolve into AGI, the metaverse would evolve into Artificial General Universe (AGU), which can be a digital twin of the entire physical universe but can also be designed to be a completely different universe with novel physical laws. When such a time comes, all the physical constraints that the intelligent agents have had to face up to now disappear or can be controlled, and accordingly intelligent agents can be freed from real-world conditions as brought by the fundamental laws of the physical universe. In this paper we elaborate upon this view and discuss social and ethical issues on the metaverse and AGU.

2 What Is AGU (Artificial General Universe)

AGU is characterized by its adaptability and freedom. In principle, it can be free not only from physical constraints (cf. [9]) but also from social constraints, such as laws, social regulations, or cultural conventions (cf. [6]). The characteristics of this environment devoid of all these constraints are the very factors that allow AGI to maximize its performance. In the following, we first discuss physical constraints and then social constraints. In the real world, many physical constraints limit the performance of intelligent agents. Agents in AGU are not necessarily constrained by location, space, or resources, which can also be easily replicated just like copying digital text. Also, they can be freed from real-world gravity or kinetic energy and can continue to exist indefinitely. All those physical factors can be controlled in any way. Machine learning algorithms generally require considerable resources for training and execution (e.g., hardware and computing power, time and location, and human staff involved). However, AGU does not have these physical constraints, allowing for intelligence scalability. The real world often places ethical, legal, and social restrictions on intelligent agents. In the real world, intelligent agents are required to follow human-made regulations and comply with human values, which however slows the process of intelligence evolution. It is indeed necessary for human well-being to place those restrictions on intelligent agents. Yet at the same time, there is no such necessity in the case of AGU, in which humans and artificial agents can live separately; or we can create an AGU just for AGI. In such a way, the process of general intelligence evolution will be significantly more accelerated. Summing up, AGI in AGU will be able to maximize its intelligence through an environment that provides the flexibility and freedom that are inherent features of AGU. AGU is a universe in which various AI/AGI agents can be accommodated as well as human agents. All this can be summarised by the following two figures (Figs. 1 and 2).

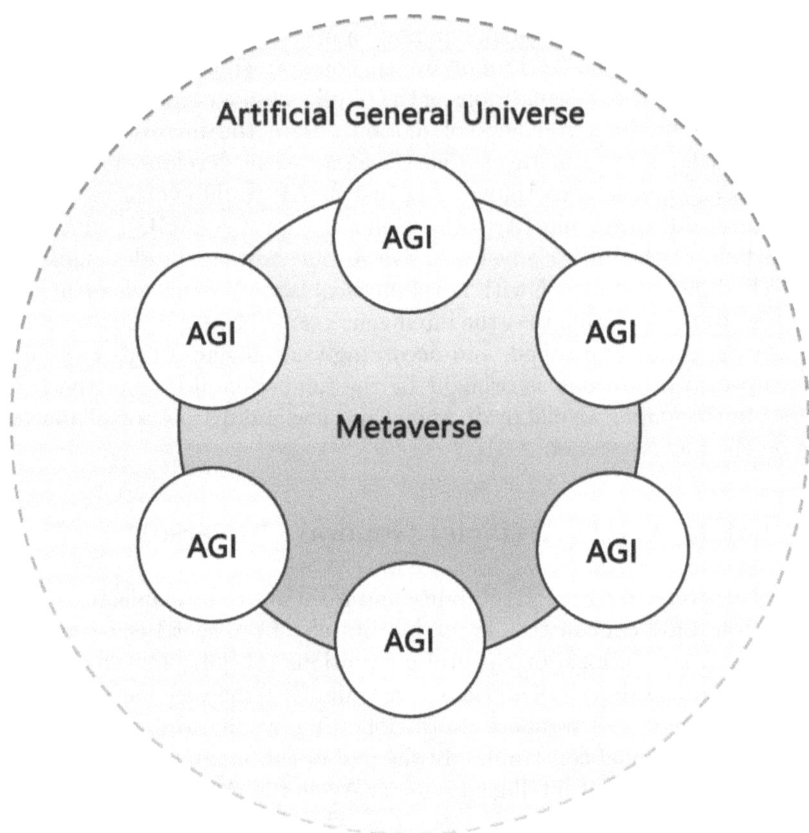

Fig. 1. The Landscape of Artificial General Universe

3 Social Issues on the Metaverse and AGU

It has been extensively discussed in AI ethics how AI and AGI would impact human society. The metaverse and AGU would have an enormous impact on human society (as well as AI/AGI society); yet at the same time, metaverse ethics is still under development. Here we shall discuss social issues on the metaverse and AGU. There are both new issues brought into society by them and old issues reinforced by them; some issues recognized in AI ethics become more severe in the metaverse and AGU.

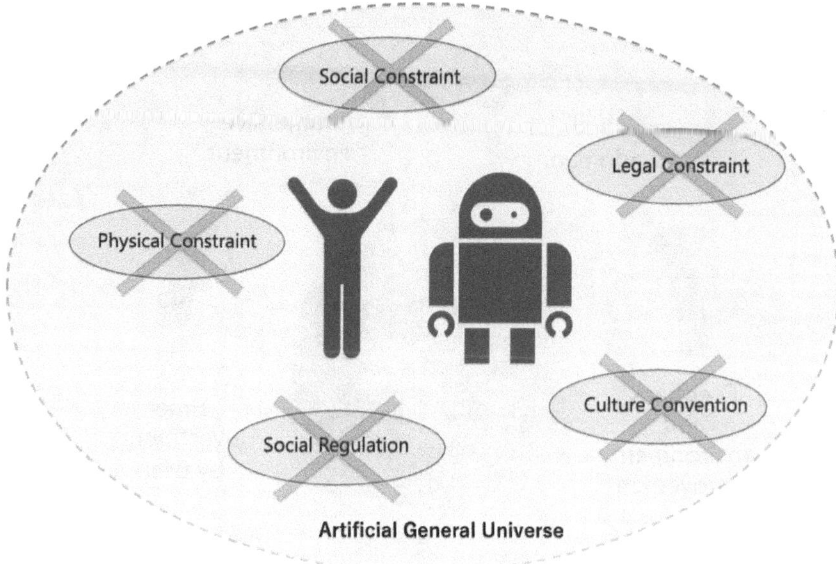

Fig. 2. Escaping from Real-World Constraints in Artificial General Universe

3.1 Surveillance Capitalism

The issue of privacy and surveillance capitalism, which is also a major topic in AI ethics, will become even more serious in AGU. This is because most aspects of our lives are informationalized in AGU, being represented as digital information easy to be monitored on a real-time basis and recorded for further analysis (cf. [5]). All sorts of information on human bodies and minds will be able to be collected in AGU, including information on sight, hearing, taste, smell, speech, body movement, and possibly even the content of thought (although the sorts of information that are already available in current metaverses are still limited). In addition, AGU is addictive and immersive (cf. [25]); therefore, users are likely to spend a significant amount of time there, thus increasing the overall amount of information collected. Thus, AGU increases the variety and quantity of information collected, which in turn increases the scale of the surveillance problem and accelerates surveillance capitalism. Gina Neff interviewed in [3] warns that this could limit the ability of regulators and the state to protect individuals and concentrate power in the hands of corporations (assuming that metaverses are developed and maintained by them; there can be more open, decentralized metaverses as well) (Fig. 3).

3.2 Issues at the Border Between the Real and Virtual

Since the immersive nature of the metaverse and AGU blur the line between reality and virtuality, it causes some issues [3]. For example, a user may participate in a virtual war there, which can damage the user mentally and physically

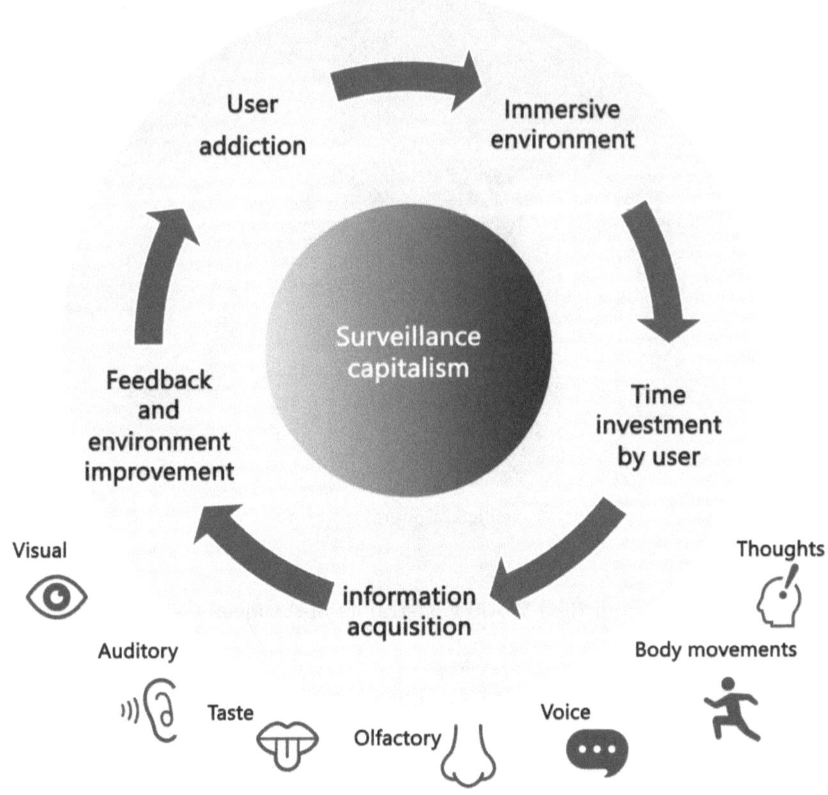

Fig. 3. Surveillance Capitalism

(especially if it goes beyond audiovisual space and allows for diverse sense perceptions); the user will be attacked more realistically than if it were simply played on a screen. Even after leaving the virtual world, the user may suffer in the real world from the damage caused by the virtual experience. Another issue is that some problems that occur in the real world are brought into the virtual environment. For example, discrimination based on race or gender occurring in the real world may be transferred to the virtual space [4]. Discrimination and bias issues have been much discussed in AI ethics; however, if they are brought into the metaverse and AGU, they would become more serious; discrimination then happens globally on a cosmic scale rather than locally on an individual agent scale. As exemplified by this, the metaverse and AGU tend to amplify the AI ethics problems that exist locally on a single agent to more global problems on a cosmic scale (Fig. 4).

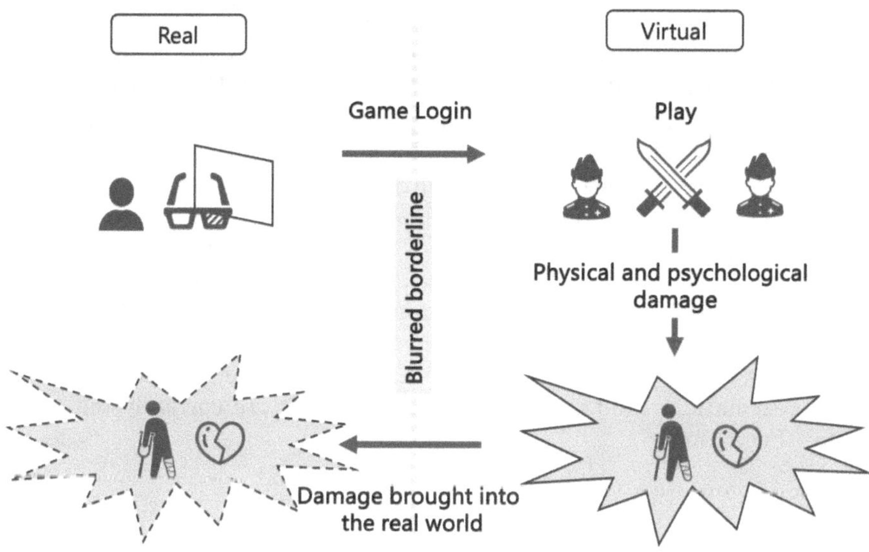

Fig. 4. The Example of the Issues because of the Blurred Borderline

3.3 Governance, Regulation, and Accountability

There is currently no clear governance or regulation on the metaverse and AGU. As information systems become more integrated into various industries and applications, questions have been raised as to who is responsible for the actions and outcomes of these systems [8]. In particular, it is not obvious who is responsible for what in the metaverse and AGU. Suppose, for example, that a human lives there as an avatar and that the avatar did something wrong. Should we punish the human in the actual universe directly or just the avatar in the artificial universe? Even the death of the avatar may mean nothing to the human that operates it; yet, considering the aforementioned issue, killing the avatar may damage the operating human (who experiences virtual death, which can be shocking enough to damage them). There may be multiple avatars operated by the same human, and in that case, punishing the operating human leads to punishing all the avatars even if only one of them did something wrong. Is it fair or not? If avatars are regarded as autonomous agents, then they should be punished individually and should not be treated collectively even if they were originally made by the same human. All this ultimately depends upon what sorts of laws and rules are implemented in the metaverse and AGU. In current society, responsibility for information systems is attributed primarily to human entities, based on the laws and rules established by humans [7]. However, AGU could be an autonomous universe (although current metaverses are not), and would even be able to design its laws and rules, serving as the governor of the artificial universe, which may still sound like a fantasy, but there are related ideas such as algorithmic politics (cf. [26]), which could improve upon human politics. Similarly, governance by AGU as an autonomous agent could be better (in a suitable sense) than governance by humans; especially considering that there are currently no truly international laws that govern the entire earth across all the different nations (and thus international conflicts are currently difficult to resolve) (Figs. 5 and 6).

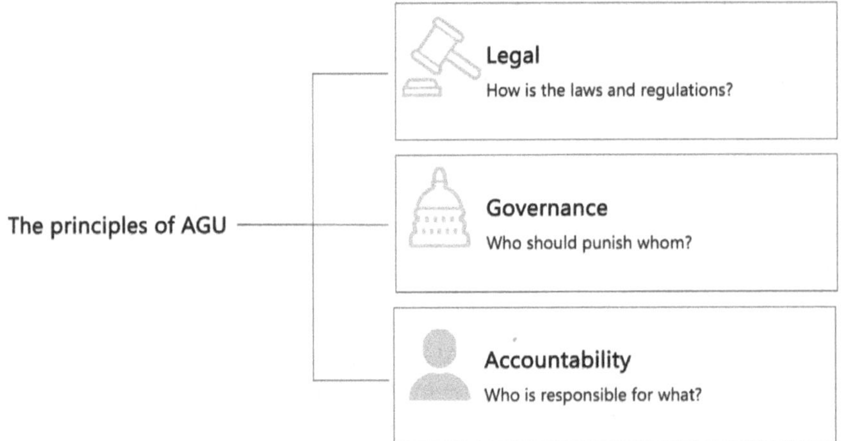

Fig. 5. The Principles of AGU

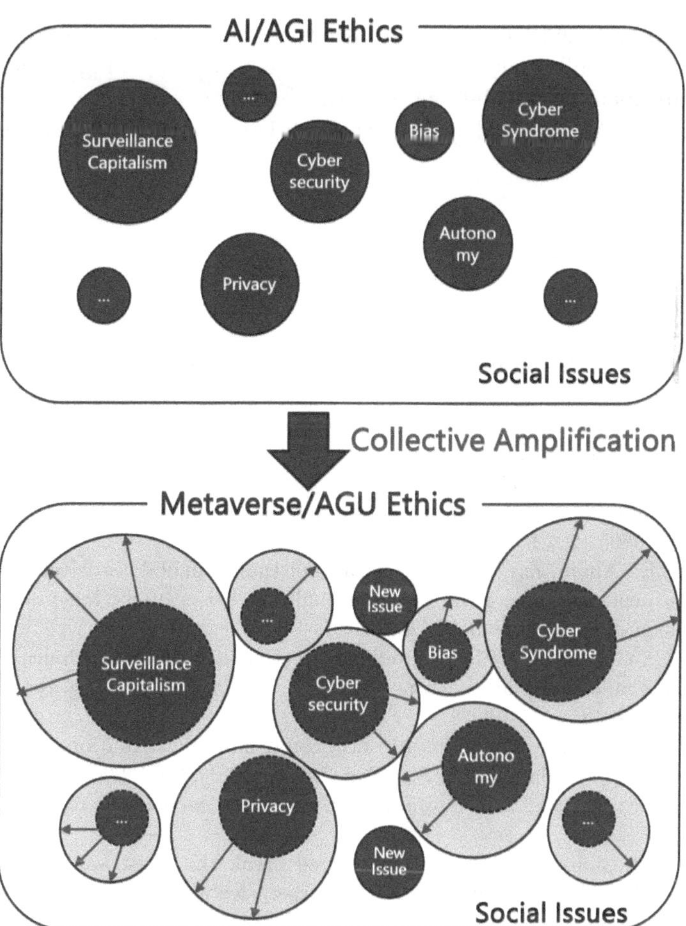

Fig. 6. The Acceleration of Social and Ethical Problems in the Metaverse/AGU

4 Concluding Remarks

We have discussed how the metaverse would transform the world and in particular how it would impact AI and AGI. AGU, the ultimate form of the metaverse, is a universe free from the physical and social constraints of the real world. Its flexibility and freedom provide an environment where the performance of AGI can be maximized; AGI could even design a suitable AGU for itself to maximize the performance. We have also discussed social issues such as the acceleration of surveillance capitalism due to the informationalization of the entire universe, the emergence of problems between the real and virtual, and new forms of governance and regulation on AGU. Comparing AI and metaverse ethics, AI ethics problems tend to be amplified in artificial universes since they grow there on a cosmic scale, which means that the social and ethical problems as discussed above would become more and more severe as the metaverse develops, which is illustrated in the following figure. It would thus be a compelling challenge of our time to implement appropriate measures against them and prevent the dystopian acceleration of them from damaging the well-being of the human race.[1]

References

1. Abbott, V., Xu, T., Maruyama, Y.: Category theory for artificial general intelligence. In: Thorisson, K.R., et al. (eds.) AGI 2024. LNAI, vol. 14951, pp. 119–129. Springer, Cham (2024)
2. Bloomfield, C., Maruyama, Y.: Fibered universal algebra for first-order logics. J. Pure Appl. Algebra **228**, 107415 (2023)
3. Anderson, J., Rainie, L.: The Metaverse in 2040, Pew Research Center: Internet, Science & Tech (2022)
4. Anshari, M., Syafrudin, M., Fitriyani, N.L., Razzaq, A.: Ethical responsibility and sustainability (ERS) development in a metaverse business model. Sustainability **14**(23), 15805 (2022)
5. Bibri, S.E., Allam, Z.: The metaverse as a virtual form of data-driven smart urbanism: on post-pandemic governance through the prism of the logic of surveillance capitalism. Smart Cities **5**(2) (2022)
6. Cath, C.: Governing artificial intelligence: ethical, legal and technical opportunities and challenges. Philos. Trans. Roy. Soc. A Math. Phys. Eng. Sci. **376**(2133), 20180080 (2018)
7. Coeckelbergh, M.: Artificial intelligence, responsibility attribution, and a relational justification of explainability. Sci. Eng. Ethics **26**(4), 2051–2068 (2020)
8. Doshi-Velez, F., et al.: Accountability of AI under the law: the role of explanation (2017)
9. Sachin, M., John, C.: Can the metaverse break the constraints of the physical world? DBS Group Research (2022). https://www.dbs.com/metaverse/can-the-metaverse-break-the-constraints-of-physical-world.html
10. Maruyama, Y.: Fundamental results for pointfree convex geometry. Ann. Pure Appl. Logic **161**, 1486–1501 (2010)
11. Maruyama, Y.: Natural duality, modality, and coalgebra. J. Pure Appl. Algebra **216**, 565–580 (2012)
12. Maruyama, Y.: From operational chu duality to coalgebraic quantum symmetry. In: Heckel, R., Milius, S. (eds.) CALCO 2013. LNCS, vol. 8089, pp. 220–235. Springer, Heidelberg (2013). https://doi.org/10.1007/978-3-642-40206-7_17
13. Maruyama, Y.: Full Lambek Hyperdoctrine: categorical semantics for first-order substructural logics. In: Libkin, L., Kohlenbach, U., de Queiroz, R. (eds.) WoLLIC 2013. LNCS, vol. 8071, pp. 211–225. Springer, Heidelberg (2013). https://doi.org/10.1007/978-3-642-39992-3_19
14. Maruyama, Y.: Categorical duality theory: with applications to domains, convexity, and the distribution monad. Leibniz Int. Proc. Inf. **23**, 500–520 (2013)

[1] To that end we have proposed and developed categorical artificial intelligence as the integration of symbolic and statistical AI for verifiable, ethical, and trustworthy AI (see [1,23,27]; see also [18–20,22,24]) while building upon category theory as the foundational language of structural integration (see, e.g., [2,10–17,21]).

15. Maruyama, Y.: Prior's Tonk, notions of logic, and levels of inconsistency: vindicating the pluralistic unity of science in the light of categorical logical positivism. Synthese **193**, 3483–3495 (2016)
16. Maruyama, Y.: Categorical harmony and paradoxes in proof-theoretic semantics. In: Piecha, T., Schroeder-Heister, P. (eds.) Advances in Proof-Theoretic Semantics. TL, vol. 43, pp. 95–114. Springer, Cham (2016). https://doi.org/10.1007/978-3-319-22686-6_6
17. Maruyama, Y.: Meaning and duality: from categorical logic to quantum physics. Ph.D. thesis, Department of Computer Science, University of Oxford (2017)
18. Maruyama, Y.: Compositionality and contextuality: the symbolic and statistical theories of meaning. In: Bella, G., Bouquet, P. (eds.) CONTEXT 2019. LNCS (LNAI), vol. 11939, pp. 161–174. Springer, Cham (2019). https://doi.org/10.1007/978-3-030-34974-5_14
19. Maruyama, Y.: Symbolic and statistical theories of cognition: towards integrated artificial intelligence. In: Cleophas, L., Massink, M. (eds.) SEFM 2020. LNCS, vol. 12524, pp. 129–146. Springer, Cham (2021). https://doi.org/10.1007/978-3-030-67220-1_11
20. Maruyama, Y.: The conditions of artificial general intelligence: logic, autonomy, resilience, integrity, morality, emotion, embodiment, and embeddedness. In: Goertzel, B., Panov, A.I., Potapov, A., Yampolskiy, R. (eds.) AGI 2020. LNCS (LNAI), vol. 12177, pp. 242–251. Springer, Cham (2020). https://doi.org/10.1007/978-3-030-52152-3_25
21. Maruyama, Y.: Fibred algebraic semantics for a variety of non-classical first-order logics and topological logical translation. J. Symbolic Log. **86**, 1189–1213 (2021)
22. Maruyama, Y.: Category theory and foundations of life science. Biosyst. J. **203**, 104376 (2021)
23. Maruyama, Y.: Categorical artificial intelligence: the integration of symbolic and statistical AI for verifiable, ethical, and trustworthy AI. In: Goertzel, B., Iklé, M., Potapov, A. (eds.) AGI 2021. LNCS (LNAI), vol. 13154, pp. 127–138. Springer, Cham (2022). https://doi.org/10.1007/978-3-030-93758-4_14
24. Maruyama, Y.: Moral philosophy of artificial general intelligence: agency and responsibility. In: Goertzel, B., Iklé, M., Potapov, A. (eds.) AGI 2021. LNCS (LNAI), vol. 13154, pp. 139–150. Springer, Cham (2022). https://doi.org/10.1007/978-3-030-93758-4_15
25. Mystakidis, S.: Metaverse. Encyclopedia **2**(1), 486–497 (2022)
26. Srivastava, S.: Algorithmic governance and the international politics of big tech. Perspect. Polit. **21**(3), 989–1000 (2023)
27. Xu, T., Maruyama, Y.: Neural string diagrams: a universal modelling language for categorical deep learning. In: Goertzel, B., Iklé, M., Potapov, A. (eds.) AGI 2021. LNCS (LNAI), vol. 13154, pp. 306–315. Springer, Cham (2022). https://doi.org/10.1007/978-3-030-93758-4_32

Author Index

K. R. Thórisson et al. (Eds.): AGI 2024, LNAI 14951, p. 239, 2024.
https://doi.org/10.1007/978-3-031-65572-2

SPRINGER NATURE

GPSR Compliance

The European Union's (EU) General Product Safety Regulation (GPSR) is a set of rules that requires consumer products to be safe and our obligations to ensure this.

If you have any concerns about our products, you can contact us on ProductSafety@springernature.com

In case Publisher is established outside the EU, the EU authorized representative is:

Springer Nature Customer Service Center GmbH
Europaplatz 3
69115 Heidelberg, Germany